NIMS Monographs

The NIMS Monographs are published by the National Institute for Materials Science (NIMS), a leading public research institute in materials science in Japan, in collaboration with Springer. The series present research results achieved by NIMS researchers through their studies on materials science as well as current scientific and technological trends in those research fields.

These monographs provide readers up-to-date and comprehensive knowledge about fundamental theories and principles of materials science as well as practical technological knowledge about materials synthesis and applications.

With their practical case studies the monographs in this series will be particularly useful to newcomers to the field of materials science and to scientists and engineers working in universities, industrial research laboratories, and public research institutes. These monographs will be also available for textbooks for graduate students.

National Institute for Materials Science
http://www.nims.go.jp/

More information about this series at http://www.springer.com/series/11599

Makoto Tachibana

Beginner's Guide to Flux Crystal Growth

Makoto Tachibana
National Institute for Materials Science
Tsukuba, Ibaraki
Japan

ISSN 2197-8891 ISSN 2197-9502 (electronic)
NIMS Monographs
ISBN 978-4-431-56586-4 ISBN 978-4-431-56587-1 (eBook)
DOI 10.1007/978-4-431-56587-1

Library of Congress Control Number: 2017952003

Printed on acid-free paper

This Springer imprint is published by Springer Nature
The registered company is Springer Japan KK
The registered company address is: Chiyoda First Bldg. East, 3-8-1 Nishi-Kanda, Chiyoda-ku, Tokyo 101-0065, Japan

Preface

This small book is written mainly for beginning graduate students and researchers in solid-state physics, who are thinking of growing single crystals in order to study their physical properties.

For students and professional scientists alike, solid-state physics offers many fascinating research topics. Superconductivity that competes and coexists with magnetic order; a new type of ferroelectricity arising at magnetic transitions; and quantum materials that conduct electriticity only along their surfaces—these are just few examples of what solid-state physicists today are trying to understand and find possible uses for. To study these phenomena, researchers perform various kinds of measurements, from simple resistivity tests to state-of-the-art experiments at multi-national synchrotron facilities. Common to all these studies, however, is the need to obtain high-quality single crystals for detailed investigation.

This book focuses on the principles and techniques of growing high-quality single crystals using the flux method. Although it is only one technique among many for growing crystals, the flux method is favored by many solid-state physicists. In this method, single crystals are obtained by cooling a hot molten liquid of the desired compound dissolved in a flux (another term for flux is solvent; flux growth is also called high-temperature solution growth). Because it is possible to find a flux for most inorganic materials, the technique can be used to obtain a wide variety of crystals having sizes of the order of several millimeters—an appropriate dimension for most physical measurements. Moreover, the flux method only requires a standard electric furnace and crucible, and does not demand too much time and effort on the part of the scientist. In other words, by using the flux method, the novice researcher can learn to grow many crystals without becoming a dedicated crystal grower.

This book assumes that the reader has some knowledge of the basic concepts of solids, such as crystal structure and chemical bonding. On the other hand, no hands-on experience in research is assumed and it is hoped that the practical approach of this book will be of help in setting up a lab and conducting successful crystal growth. Since oxides represent one of the most widely studied groups of compounds, many examples in this book come from experiments on oxide systems.

Nevertheless the flux technique can be equally applied to other materials, and this point is emphasized in various parts of the book.

I would like to thank the editors and reviewers for many suggestions which improved the quality of this book. My appreciation also goes to all those who have helped me in my crystal growth activities over the years. Most of the illustrations in this book were prepared by Rie Tachibana and Marisa Tachibana.

Tsukuba, Japan Makoto Tachibana

Contents

Chapter 1
Introduction

1.1 Single Crystals in Solid-State Research

This is a book about flux crystal growth, written for anyone who wants to obtain crystalline samples for physical studies. In this opening section, the merits of growing crystals are presented to those readers with no experience in solid-state research. First, I would like to recount my own experience:

I was introduced to the field of crystal growth by luck. My main interest, at the beginning of my graduate studies, was in the accurate determination of the physical properties of solids. I had chosen my thesis advisor accordingly, and I was learning to use a custom-built calorimeter for heat capacity measurements on various solids. As the samples came from the professor's collaborators, I did not really have a good idea of how they were made or why I should be studying them. I was just happy that I was making measurements and analyzing data.

Then, during my second year, I read several papers on a compound ($Cd_2Nb_2O_7$) showing intriguing ferroelectric behavior. It was clear from the papers that different groups of authors disagreed on the origin of the ferroelectric properties, and I thought heat capacity measurements will immediately resolve the controversy. A problem was that my professor did not know anyone making the compound. Since part of the controversy was due to variations in sample quality, it was essential for me to perform measurements on a high-quality sample, preferably single crystals. Could I make the crystals myself? One of the papers mentioned that single crystals could be grown by the flux method, using slow cooling from 1250 °C—a temperature easily attained in one of the electric furnaces in the laboratory. With the help of a senior graduate student, I was able to grow beautiful crystals in a short time. The crystals led to an interesting heat capacity study, and the results helped to clarify the nature of the ferroelectric behavior [1].

This episode is provided because it shows that growing single crystals: (1) does not have to be difficult; (2) enables you to determine your own research topic; and

M. Tachibana, *Beginner's Guide to Flux Crystal Growth*, NIMS Monographs,
DOI 10.1007/978-4-431-56587-1_1

(3) gives you a good idea of the origin, and also quality, of your research sample. Of course, not all crystals are easy to grow, and many solid-state physicists succeed in their careers without ever growing a crystal. Nevertheless, many physicists would agree that having the ability to grow crystals will give you the power to conduct original research [2, 3], even if you are studying basic properties such as heat capacity or resistivity. This viewpoint will be explored in this section.

1.1.1 What Is a Single Crystal?

As we have already used the words "single crystals", this is a good point to examine this term more closely. Many scientists would define a crystal as a solid, composed of atoms arranged in an orderly, repetitive array. This is a perfectly valid definition, and most solids that are not amorphous (like glasses) are indeed crystals. However, most crystalline objects are not one chunk of a crystal, but instead are "polycrystals" made up of many crystals. A sheet of aluminum foil is made up of millions of crystals of aluminum, rather than a single crystal. Similarly, a piece of magnet is composed of many crystals of a ferromagnetic material. These objects are, therefore, what we call polycrystals or polycrystalline materials. Compared with polycrystals, we see fewer products of single crystals in our daily lives; perhaps the nearest example is synthetic sapphire (single-crystalline Al_2O_3), which is used as glass for high-quality watches and as lens covers and touch screens on some smartphones.

Figure 1.1 is a photograph of the rather complex compounds $SrBi_2Ta_2O_9$ (abbreviated as SBTO) and $Bi_2Sr_2CaCu_2O_{8+x}$ (abbreviated as BSCCO, pronounced "bisko"). SBTO is a material used in ferroelectric memories. BSCCO is a metal that transforms into a superconductor below $\sim$85 K. For each compound, single crystals are shown at the bottom of the photograph. The single crystals of SBTO are transparent, because the compound is a good electrical insulator and most rays of visible light pass through the crystals. On the other hand, the metallic BSCCO both reflects and absorbs light and is consequently opaque. The single crystals of both compounds are shiny because some light is reflected off the flat faces of the crystals. Both crystals are also very thin and can be cleaved (separated) easily along their flat faces—these properties are due to the layered crystal structure, with the large flat faces being parallel to the layer direction. The electrical resistivity of BSCCO at room temperature is much higher across the layers than along the layers.

For both SBTO and BSCCO, polycrystalline samples are also shown in the photograph, placed above the single crystals. As in aluminum foil and pieces of magnet, these polycrystals are made up of many small crystals. Indeed, observation under a microscope would reveal the presence of many micrometer-sized crystals, called crystallites, randomly oriented with respect to each other. These crystallites are also called grains, and the regions between crystallites are called grain

Fig. 1.1 Bottom: Single crystals of $SrBi_2Ta_2O_9$ (left) and $Bi_2Sr_2CaCu_2O_{8+x}$ (right). Top: Polycrystalline pieces of the same compounds. (mm grid)

boundaries. Because light is scattered at voids along the grain boundaries, the polycrystalline samples of SBTO are opaque. Electrical currents are also scattered at grain boundaries. This, combined with the random distribution of the layer structure, makes the resistivity of BSCCO polycrystals much higher than that of the single crystals along the layer direction. Owing to these extrinsic effects, many types of measurements can be performed more accurately on single crystals.

Although single crystals are usually preferred in solid-state research, there are many instances where polycrystalline samples are used to study physical properties. For some compounds, single crystals have not been prepared in a pure form or in any form at all. Some properties can be measured equally well on single crystals and polycrystals, and it is usually much easier to prepare polycrystals.

In many cases, the properties of a new compound are first studied using polycrystals. The structural parameters, as well as electrical, magnetic, and thermal properties, are often the first to be investigated. These provide an approximate, but good, idea of the compound. If the properties seem interesting, or if there is a good chance of new physics hidden in the compound, scientists may attempt to grow single crystals. The quality and size of single crystals usually improve as more scientists attempt different growth experiments. The best crystals have minimum defects and impurities, offering a better understanding of the compound; these crystals are often sent to various scientists who eagerly perform special measurements within their area of expertise. There are many properties to be examined, so a full understanding of the compound may take years of research by different scientists (since each result is published as a research paper, scientists can easily grasp the extent of current understanding). Sometimes totally unexpected results emerge when better crystals are studied, new measurement techniques are used, or different ideas are tested—these also happen to old compounds that were long forgotten or believed to be fully understood. In each case, progress depends on the availability of high-quality crystals, and this is why crystal growth is very important in solid-state research.

1.1.2 Appropriate Size of Crystals for Solid-State Research

Now that we know the importance of single crystals, it is also important to have an idea of the size of crystals needed for solid-state research. After all, even the best crystals are of little value if they are too small to be used in measurements. To pursue the question of appropriate crystal size, let us first take a look at modern measurement equipment. Figure 1.2a is a photograph of the physical property measurement system (PPMS), a commercial apparatus produced by Quantum Design. The PPMS allows scientists to measure various physical properties of solids, such as resistivity, the Hall effect, the Seebeck effect, thermal conductivity, magnetic susceptibility, and heat capacity. The object on the left in the photograph is a cryostat, into which liquid helium is filled to cool the sample inside; the temperature can be continuously varied from 1.8 to 400 K. A superconducting magnet is also placed inside the cryostat, which allows the magnetic field to be varied between −7 and 7 T. (There are options to extend the temperature down to 0.05 K and the field up to 16 T.) The computer and electronic controllers on the right-hand side allow measurements to be carried out without human intervention, so the operator is free to leave the room once the sample is set inside the PPMS. A typical set of measurements usually takes somewhere between a few hours and a few days.

Some scientists in academia are not particularly fond of the PPMS, because it allows students to obtain data without understanding the underlying concepts of measurements. The PPMS is also expensive and not all scientists who wish to use it have easy access. Nevertheless, the PPMS and other commercial apparatuses are becoming a part of typical scenes in laboratories, because they allow scientists to perform various kinds of measurements without becoming an expert in each technique. Today, it is not difficult for a solid-state physicist to conceive a research plan, grow and characterize single crystals, perform many measurements on the crystals, and write up a paper all by himself or herself. This is very different from some branches of physics (such as high-energy particle physics), where hundreds or even thousands of scientists cooperate to publish a single paper.

We are now widely deviating. Getting back to the question of appropriate crystal size, Fig. 1.2b shows a resistivity puck on the left and a heat capacity puck on the

(a)

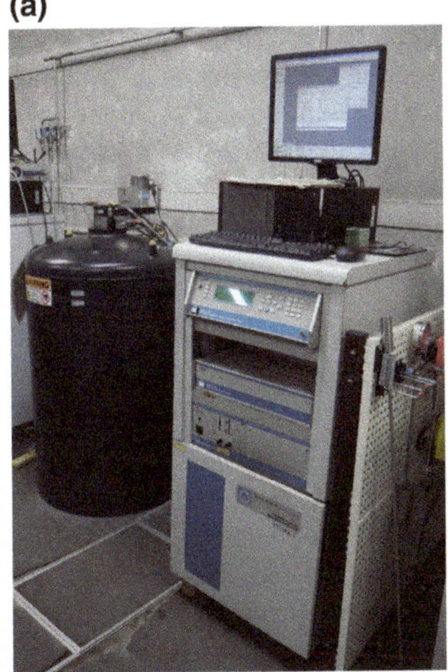

(b)

Fig. 1.2 **a** Physical property measurement system (PPMS). **b** Resistivity puck (left) and heat capacity puck (right) used with the PPMS. Each puck is about 24 mm across. The thermal shield cap for the heat capacity puck is also shown

right, which are used with the PPMS. The diameter of each puck is about 24 mm. "Puck" is just a cute name for the sample attachment device, which is inserted into the cryostat using a special stick. For resistivity measurements, four wires are connected between the sample and the puck; the outer two wires apply currents to the sample, and the inner two wires measure the voltage drop. For heat capacity measurements, the sample is attached to the platform of the puck using cryogenic grease. Heat capacity is measured by applying a heat pulse to the sample and monitoring the decrease in temperature once the heat pulse is turned off. Other types of measurements in the PPMS are performed using different pucks or inserts.

Figure 1.2b provides a good idea of the size of crystals needed for these measurements. The typical weight of a sample for a heat capacity measurement is 30 mg, with a minimum acceptable weight being about 1 mg. If the sample is much lighter than 1 mg, the relatively large heat capacities of the platform and grease will mask the sample's small heat capacity. A cube of copper with a side measuring 1.5 mm weighs 30 mg, so this is a good size for heat capacity measurements. The resistivity measurement, on the other hand, does not have a minimum acceptable size. However, it is very difficult to attach four wires to a small crystal, and many scientists would prefer to work with crystals that have sides measuring at least 1 mm. Many other measurements using the PPMS, as well as those not included in the PPMS (such as various optical and resonance measurements), also work well with samples of similar dimensions. Therefore, it is convenient for our purpose to set an appropriate size of crystal larger than 1 mm, in at least one direction but preferably in two or all three directions. An important exception to this general remark is during measurements of neutron inelastic scattering, which usually requires crystals of 1 cm^3 or larger. Single-crystalline thin films of compounds can also be used for some types of measurements, but they are usually studied by slightly different groups of scientists.

1.1.3 Types of Crystals Frequently Studied in Solid-State Physics

So far, we have looked at what single crystals are and how big they have to be for physical measurements. So, the next question is: Which groups of compounds have received the most interest from solid-state physicists in recent years? (Let us recall that there are more than 100 elements in the periodic table, shown in Fig. 1.3, capable of combining to form an inexhaustible variety of compounds.) While the question does not have a simple objective answer, it may be answered by listing some of the popular fields of research in the recent past, as judged by the number of news articles that have appeared in *Nature*, *Science*, and *Physics Today*. Solid-state physics is often referred to as condensed-matter physics in these periodicals. Below is my list, which covers the last 30 years and includes the most typical compounds studied in each field. Many other compounds have been studied within the context of these fields.

1. High-temperature superconductivity in copper oxides ($La_{2-x}Sr_xCuO_4$, $YBa_2Cu_3O_7$).
2. Colossal magnetoresistance in manganese oxides ($La_{1-x}Sr_xMnO_3$, $La_{1-x}Ca_xMnO_3$).
3. Heavy-fermion superconductors ($CeCoIn_5$, URu_2Si_2).
4. Low-dimensional quantum magnetism and geometrically frustrated magnetism ($CuGeO_3$, $Dy_2Ti_2O_7$).
5. Relaxor ferroelectrics ($PbMg_{1/3}Nb_{2/3}O_3$).
6. Ferroelectric magnets or "multiferroics" ($TbMnO_3$, $BiFeO_3$).
7. High-temperature superconductivity in iron-based compounds ($BaFe_2As_2$, FeSe).
8. Topological insulators (Bi_2Se_3, SmB_6).

This list is far from complete and only those fields in which single crystals of inorganic and non-molecular compounds are studied are included. Moreover, the list excludes those fields in which traditional "textbook" compounds such as GaAs and NiO are studied, and it is to be remembered that the list shows what has been actively studied in the recent past, rather than what will be fashionable in the future. What all these fields of research mean is not of concern here, but what is important are the types of compounds making up the list. Transition-metal oxides are well

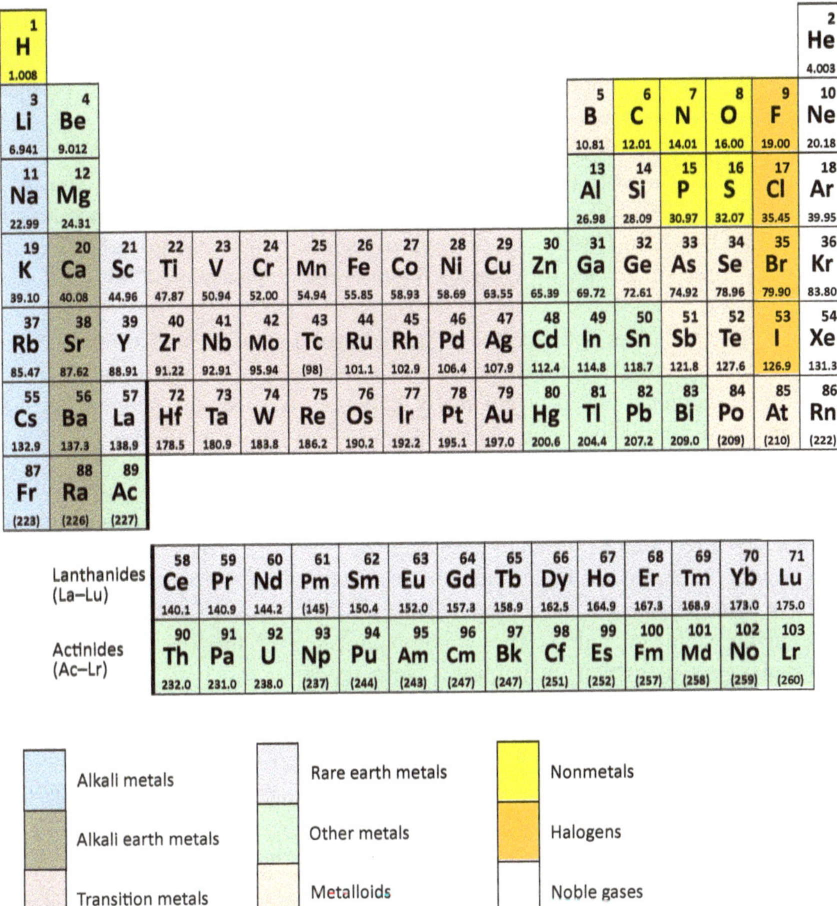

Fig. 1.3 Periodic table of the elements. The classification for several elements is up for some debate

represented, followed by intermetallic compounds (materials made solely of metals and metalloids). $La_{1-x}Sr_xMnO_3$, $La_{1-x}Ca_xMnO_3$, $PbMg_{1/3}Nb_{2/3}O_3$, $TbMnO_3$, and $BiFeO_3$ all belong to the same structural group and are called perovskite oxides. Many compounds in the list are ternary compounds, meaning they are composed of three different elements at distinct structural sites. Compounds such as $La_{2-x}Sr_xCuO_4$ and $La_{1-x}Sr_xMnO_3$ are pseudo-ternary, because they are solid solutions where the La^{3+} and Sr^{2+} ions are randomly placed in the same structural site. (The Sr^{2+} ions are also called dopants, especially when the value of x is small.) With the oxygen ions being O^{2-}, the oxidation state, or valence, of the transition-metal ions varies with x to keep the compounds electrically neutral as a whole. The value of x strongly controls the electronic properties of these compounds, with the superconductivity and colossal magnetoresistance appearing only in a limited range of x.

Perhaps the most important feature of the list is the fact that most of these compounds were known long before the fields became popular in solid-state physics. For example, Bi_2Se_3 had been studied since the 1950s as a thermoelectric material, before it was recognized as a topological insulator in 2009 [4]. Similarly, $BiFeO_3$, $PbMg_{1/3}Nb_{2/3}O_3$, and FeSe were already known as stable compounds in the 1950s. Only $YBa_2Cu_3O_7$ was discovered as a truly new compound within the context of these studies [5, 6], whereas the rest were "rediscovered" by solid-state physicists.

The above remark is not meant as an insult to solid-state physicists; rather, it just shows that physicists are most interested in searching and understanding novel physical phenomena, and it is not really important whether the phenomena take place in a new compound or an old compound. With this in mind, it is possible to divide the crystal growth activity into four categories:

1. Growing single crystals of known compounds by following the procedures (or "recipes") reported in the literature.
2. Growing larger and/or higher quality single crystals of known compounds that have been prepared as single crystals.
3. Growing single crystals of known compounds for the first time.
4. Growing single crystals of unknown compounds.

The difficulty and exploratory nature of the activity usually increases from 1 to 4, but, as far as solid-state physics is concerned, all activities are important for advancing our understanding of solids. In this book, all four types of activity will be considered.

1.1.4 Single Crystals in Other Fields of Study

Finally, although this book is written mainly for researchers at the beginning of their careers in solid-state physics, it is worth looking into the role of crystal growth in other fields of physical science. Here, we consider solid-state chemistry, applied physics, and mineralogy.

Solid-state chemistry is closely related to solid-state physics. Perhaps the most important difference is the tendency of solid-state chemists to focus on new compounds with unique or complicated crystal structures, whereas solid-state physicists usually prefer to study complex phenomena in structurally simple systems. Because crystal structures can be studied adequately with polycrystals or tiny single crystals, the growth of large (>1 mm^3) single crystals is not often the major concern of solid-state chemists. It is usually left to the solid-state physicist to recognize interesting compounds and to grow single crystals that are sufficiently large for physical measurements.

Applied physicists use the knowledge of fundamental physics to conceive and build practical devices for applications. Many topics in solid-state physics—such as superconductivity, magnetism, ferroelectricity, and semiconductor physics—are not only intellectually stimulating but also extremely useful. An applied physicist would grow single crystals of GaN, for example, to build better devices for laser diodes and high-power electronics.[1] Because making new or better devices usually requires more resources and effort than making samples for basic physical measurements, applied physicists often focus on the growth of one particular material for many years—the crystal growth of silicon for electronics is still being studied after more than 60 years. Many applied physicists are also interested in the details of how crystals grow, because the quality of devices depends so much on the purity and perfection of the crystals.

Minerals are naturally occurring crystals; therefore, man-made crystals are not minerals. Nevertheless, some mineralogists are interested in growing synthetic counterparts of minerals, because they are usually purer and provide better understanding of the various mineralogical properties. Conversely, solid-state physicists can find interesting physics in natural mineral specimens. A physicist once purchased crystals of azurite ($Cu_3(CO_3)_2(OH)_2$) at a mineral shop and decided to study its magnetic properties. It turned out that azurite is a remarkable example of a novel quantum magnet, closely following the theoretical predictions of the frustrating diamond spin-chain model [7]. The recent study of the mineral kawazulite as a topological insulator is another interesting example [8].

1.2 Overview of Flux Growth

The previous section looked at the roles of single crystals and crystal growth in physical research, without paying particular attention to the flux technique. An overview of flux growth is now provided in this section, using the growth of ruby crystals as an example. Each aspect of what is described in this section will be explored more fully in later chapters.

Ruby is composed of Al_2O_3 with about 1% of the Al^{3+} ions replaced by Cr^{3+} ions, so that its chemical formula can be written as $Al_{2-x}Cr_xO_3$ ($x \approx 0.02$). The addition of Cr^{3+} ions transforms the colorless Al_2O_3 into the intense red of ruby; natural ruby is a precious gemstone, whereas man-made ruby with a smaller concentration of Cr^{3+} was used in the first laser [9]. Using the flux method, ruby crystals can be grown by using a combination of PbO and B_2O_3 as a flux. A flux is also called a solvent, and it is used to dissolve a solute in a solution; therefore, $Al_{2-x}Cr_xO_3$ is the solute in this case.

In the flux growth experiment, mixed dry powders of Al_2O_3, Cr_2O_3, PbO, and B_2O_3, in the weight ratio of 12:0.3:90:10 [10], are placed into a platinum crucible

[1] Single crystals of GaN can be grown in a pressurized nitrogen atmosphere using Na as a flux.

with a lid. The sealed crucible is then placed into an electric resistance furnace and heated to 1250 °C. Either a box furnace or a vertical tube furnace can be used (Fig. 1.4). The crucible is kept at 1250 °C for 10 h, during which the content melts completely and becomes a uniform solution. The temperature is then lowered at a rate of 2 °C per hour, using the programmable controller of the furnace. The solute becomes less soluble as the temperature drops, eventually reaching a point of supersaturation—a condition where the concentration of the solute exceeds the solubility of the solution. Then, a number of microscopic nuclei of ruby begin to form in the solution. With a further decrease in temperature, additional solute particles are attached to the nuclei, layer upon layer, in an orderly array, eventually growing into visible crystals. After the temperature reaches 950 °C, the furnace is shut off and the crucible is cooled to room temperature. Once the crucible is removed from the furnace, the solidified flux is dissolved in hot dilute nitric acid for several days. The grown crystals of ruby, after about half of the flux has been removed, are shown in Fig. 1.5.

In this experiment, a combination of PbO and B_2O_3 is chosen as the flux because:

1. it has a low melting point of about 500 °C (separately, PbO has a melting point of 886 °C and B_2O_3 turns into a thick liquid above 450 °C);
2. it dissolves a large amount of ruby at high temperatures, and the solubility decreases substantially with decreasing temperature;
3. it does not react with Al_2O_3 or Cr_2O_3 to form a stable compound during the growth;

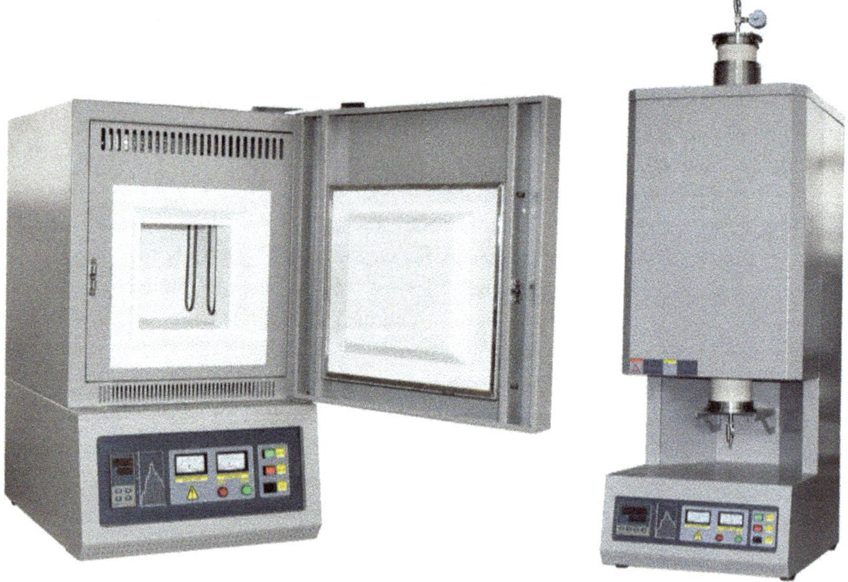

Fig. 1.4 Box furnace (left) and vertical tube furnace (right). Courtesy of Kejia Furnace

Fig. 1.5 Ruby crystals grown in a 15-ml platinum crucible. The largest crystal is about 15 mm across. The yellow regions inside the crystals are flux inclusions

4. Pb^{2+} and B^{3+} ions do not replace the Al^{3+}/Cr^{3+} ions in ruby, owing to the large differences in ionic radius (for Pb^{2+} and B^{3+}) and the difference in electrical charge (for Pb^{2+});
5. it does not attack the platinum crucible;
6. it can be easily removed from the crucible; and
7. it is readily available in high purity at a low cost.

These features make it a good flux for growing ruby crystals. On the other hand, it is not a perfect flux because:

1. B_2O_3 has a high viscosity;
2. PbO is volatile at the growth temperatures; and
3. PbO is toxic and its vapor attacks many materials.

A viscous flux slows the diffusive motion of dissolved particles in the solution, increasing the growth time and the chance of flux being trapped into the growing crystal. (See Fig. 1.5 for the presence of flux inclusions in the ruby crystals.) Therefore, the concentration of B_2O_3 in the solution should not be too high. On the other hand, the presence of B_2O_3 is important because it reduces the volatility of PbO and it also improves the size and shape of ruby crystals. Because PbO is toxic, a proper mask must be worn during the experiment to prevent accidental inhalation of its fumes and fine powders. Also, the part of the furnace that comes in contact with the PbO vapor should be replaceable. For a vertical tube furnace, this can be realized by simply replacing the ceramic tube.

Instead of slowly cooling the solution, ruby crystals can also be grown by evaporating a very volatile flux (such as PbF_2 or MoO_3) at constant temperature, or by placing a small seed crystal in the crucible and providing a proper temperature gradient. The evaporation method can reduce inhomogeneities in crystals arising from temperature variations, but it is difficult to control the growth rate using this technique. The seed technique can produce large crystals of good quality, but the experiment requires much effort in optimizing growth conditions. Aside from a brief discussion of these techniques in later chapters, much of this book will focus on the conventional slow-cooling technique without the use of a seed.

It should be pointed out that ruby crystals can be grown by carefully solidifying its own melt, without using a flux; indeed, this is how large crystals of ruby and sapphire (pure Al_2O_3) are grown in the factory. However, the melting point of ruby is about 2050 °C, a temperature too high for most furnaces and crucibles. Therefore, the primary role of a flux is to reduce the crystallization temperature, in this case from 2050 °C to below 1250 °C. The latter value is within the range of many standard furnaces, and it also allows the use of platinum crucibles (platinum has a melting point of 1768 °C and can be used with a PbO-based flux below 1350 °C).

For a compound that, before reaching its melting point, (1) decomposes, (2) transforms into another crystal structure, or (3) becomes very volatile, using a flux to lower the crystallization temperature is an effective way—sometimes the only way—to grow single crystals. A compound that decomposes into a liquid and another solid on heating is called an incongruently melting compound, the crystals of which cannot usually be grown from its own melt. Many compounds of interest to solid-state physicists melt incongruently; examples from the list in the previous section include $La_{2-x}Sr_xCuO_4$, $YBa_2Cu_3O_7$, and $BiFeO_3$.

Another advantage of the flux method is the well-formed and high-quality crystals that can be obtained by this technique. Because the crystals grow almost freely in solution without strong temperature gradients, they often show natural growth faces. This, together with relatively low temperatures and slow growth rates, often results in crystals with fewer defects and less strain than those grown from their own melts. On the other hand, it is difficult to grow very large crystals using the flux method, and contamination from the flux and crucible material is often a source of concern. In many cases, the severity of these problems can be greatly reduced by choosing optimum flux and growth conditions.

Another interesting feature of flux growth is the possibility of growing completely unexpected compounds, especially during exploratory growth experiments. As an example, scientists were trying to grow single crystals of $NpPd_3$ in a ceramic alumina (Al_2O_3) crucible using Pb as the flux. In addition to $NpPd_3$, they found crystals of a new compound, $NpPd_5Al_2$, in the product—the Al came from the partially dissolved crucible. It turned out that $NpPd_5Al_2$ is an unconventional heavy fermion superconductor with remarkable electronic properties [11].

Table 1.1 shows the typical fluxes used to grow the compounds mentioned in Sect. 1.1. (In cases where the crystals cannot be obtained by the flux method, a related compound is shown instead.) The table shows that there are similarities in chemistry between the flux and the crystal; oxide fluxes are used mostly to grow oxide crystals, whereas metallic fluxes are often used for the growth of intermetallic compounds. In some cases, a constituent of the crystals is used as the flux. For example, $La_{2-x}Sr_xCuO_4$ and $CuGeO_3$ can be grown by using an excess amount of

Table 1.1 Examples of flux-grown crystals

Crystal	Flux	T_{max} (°C)[a]	Refs.
$La_{2-x}Sr_xCuO_4$	CuO	1250	[13]
$YBa_2Cu_3O_7$	BaO–CuO	1015	[12]
$La_{1-x}Pb_xMnO_3$	$PbO–PbF_2$	1050	[14]
$CeCoIn_5$	In	1150	[48]
URu_2Si_2	In	1400	[49]
$CuGeO_3$	CuO	1220	[50]
$Dy_2Ti_2O_7$	$PbO–PbF_2–MoO_3$	1250	[51]
$PbMg_{1/3}Nb_{2/3}O_3$	$PbO–B_2O_3$	1090	[52]
$TbMn_2O_5$	$PbO–PbF_2–B_2O_3$	1280	[53]
$BiFeO_3$	Bi_2O_3	1000	[54]
$BaFe_2As_2$	FeAs	1180	[55]
FeSe	NaCl–KCl	850	[55]
Bi_2Se_3	Bi	700	[56]
SmB_6	Al	1300	[57]

[a]Starting temperature of slow cooling

CuO, while excess In is used to grow $CeCoIn_5$ crystals. CuO and In are each called a self-flux in these cases.

In the remaining sections of this chapter, other techniques of crystal growth and the literature on flux growth are briefly described. Then, the following chapters will explore various aspects of flux growth in more detail.

In Chap. 2, basic ideas on the mechanisms of crystal growth from a fluxed solution are introduced. The objective of this chapter is to provide the main ideas that are useful in growing high-quality crystals. Instead of mathematical models and equations, diagrams and photographs of crystals are used to explain some of the theoretical aspects of flux growth. A few examples of imperfections found in flux-grown crystals are also discussed in this chapter.

Chapter 3 introduces the key ideas of phase diagrams for flux growth. Knowledge of the appropriate starting composition and growth temperatures is crucial in any successful growth experiment, and such information is included in phase diagrams. Techniques for determining new phase diagrams are also described.

Chapter 4 discusses the properties and characteristics of various fluxes, as well as some tips on choosing an appropriate flux. Lists of commonly used fluxes are also provided in this chapter.

In Chap. 5, both the equipment and standard procedures of flux growth experiments are described. As slightly different techniques are involved in the growth of oxides and intermetallic compounds, they are explained separately. Discussions on choosing an appropriate furnace and crucibles are also provided. Some tips on handling and assessing the grown crystals are described at the end of this chapter.

Finally, Chap. 6 presents photographs of various crystals grown by the author using the flux technique. A total of 12 well-known compounds are chosen for this chapter; these compounds are interesting for different reasons, and in each case, the growth conditions of high-quality crystals are well described in the literature. Each photograph is accompanied by brief comments on the compound and its place in solid-state research. It is hoped that this chapter will serve as encouragement to the beginner in flux crystal growth.

The Appendix provides a list of flux-grown compounds that have been published in *Journal of Crystal Growth* since 1975.

1.3 Other Methods of Crystal Growth

Although the flux method can be used to grow single crystals of many types of compounds, it is certainly not the only technique used in solid-state research—there are compounds that have not been obtained as single crystals by the flux method, and there are compounds for which other techniques yield better crystals. To take the compounds mentioned in Sect. 1.1 as examples, the best crystals of $YBa_2Cu_3O_7$ are obtained by the flux method [12]; although single crystals of $La_{2-x}Sr_xCuO_4$ can be grown by the flux method, larger and higher quality crystals have been obtained using other techniques [13]; single crystals of $La_{1-x}Sr_xMnO_3$ have not been grown by the flux method, although excellent crystals of $La_{1-x}Pb_xMnO_3$ with similar properties can be grown by the flux method [14].

In order to put flux growth into a wider context, this section provides brief descriptions of other crystal growth techniques used in physical research. The aspiring flux grower should aim to have at least a nodding acquaintance with these techniques, and textbooks and handbooks on crystal growth should be consulted for further details. One technique missing in this section is crystal growth from the solid state. This technique, which is variously called solid-state growth, grain growth, recrystallization, or strain annealing, is carried out by keeping a polycrystalline mass slightly below its melting point. For simple metals such as elemental metals, the grains often become sufficiently large that single crystals can be cut out from the matrix. Although this technique usually fails to produce adequate single crystals for more complex compounds, the resulting small crystals can often be used for single-crystal X-ray structural studies.

1.3.1 Melt Growth Techniques

If incongruent melting, phase transformation below melting, and strong volatility are not encountered, crystals can be obtained by cooling of its own melt. However, simply solidifying the melt in a crucible usually leads to a polycrystalline mass, owing to random nucleation and uncontrolled growth. To obtain a single crystal from the melt, the solidification process is usually controlled by one of the following techniques.

(1) In the Bridgman method (Fig. 1.6a), a crucible containing the molten charge is gradually lowered from a hot zone to a cold zone of the furnace. When the pointed tip of the crucible is cooled below the melting point, one or several crystals form and continue to grow as more melt is cooled. If the crucible is sealed, this technique can be used for volatile materials.

(2) In the Czochralski method (Fig. 1.6b), a seed, which is usually a single crystal of the same compound, is attached at the tip of a drawing shaft and is dipped into a crucible containing the melt. A crystal grows from the melt as the rotating shaft is withdrawn. By carefully adjusting the growth conditions, a high-quality

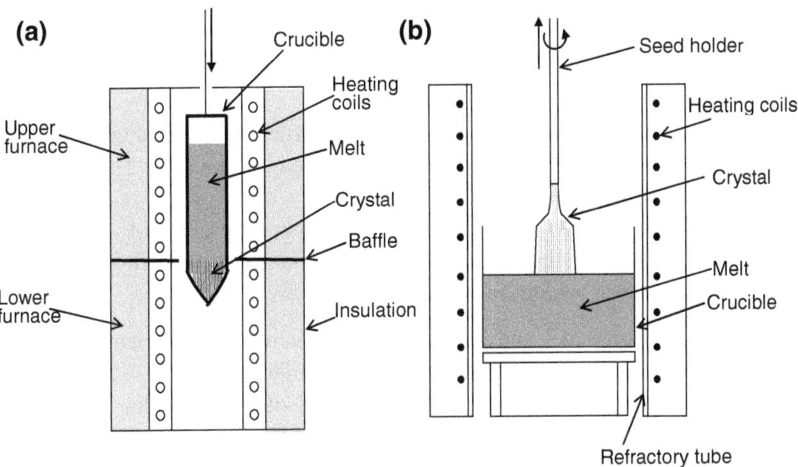

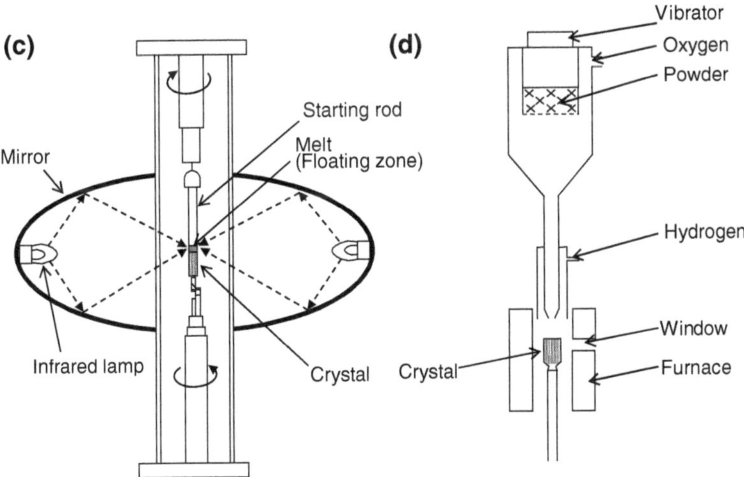

Fig. 1.6 Schematic of growth equipment for **a** Bridgman method, **b** Czochralski method, **c** optical floating zone method, and **d** Verneuil method

single crystal in the form of a rod is produced. In the tri-arc variant of this technique, three arcs are used to melt an electrically conducting material that is contained in a water-cooled copper hearth. Because the melt in this case is surrounded by a polycrystalline shell of the same substance, the tri-arc technique is useful when the melt reacts with crucible materials.

(3) In the optical floating zone method (Fig. 1.6c), ellipsoidal mirrors are used to focus the infrared light from halogen or xenon lamps onto a rotating polycrystalline rod. A single crystal is then grown by moving up the molten zone of the rod. The advantages of this technique include the absence of a crucible (which may contaminate the crystal) and the ease with which the growth atmosphere can be controlled. Also, impurities in the original rod can be swept up with the molten zone, refining the resulting crystal. For materials that do not absorb light in the infrared region, induction or an electron beam can be used as the source of heating.

(4) In the Verneuil method (Fig. 1.6d), fine powder is fed through an oxygen–hydrogen flame and is deposited onto the molten tip of a seed. This technique is suited for refractory (high-melting) oxides, but the crystals are usually highly strained from the rapid growth and steep temperature gradients. Although this technique has mostly been overtaken by the floating zone method in solid-state research, it is still widely used in industry.

Bridgman, Czochralski, and Verneuil are the names of the scientists who used the technique to grow single crystals. The Bridgman technique is also called the Bridgman–Stockbarger technique. Bridgman is too often misspelled as Bridgeman. Czochralski and Verneuil are misspelled less frequently, presumably because the correct spelling is checked more carefully.

1.3.2 Solution Growth Techniques

In addition to the flux method, there are several types of solution growth techniques that are used in solid-state research.

(1) In the aqueous solution method, water is used as the solvent. A typical temperature for growth is between room temperature and 60 °C, at which water evaporates slowly from the solution. The advantages of this method are the low cost of the apparatus and the possibility of controlling the growth conditions to a high degree. Many protein crystals have been grown by this method in recent years. Not many compounds of interest to solid-state physicists are soluble in water, but some quantum magnets have been obtained by this technique [15]. Organic liquids are also used as solvents, with a similar experimental setup.

(2) In the hydrothermal method, water is heated under pressure to raise the boiling point, under which many oxides become soluble. Mineralizers such as NaOH and Na_2CO_3 are often added to increase solubility. Typical growth conditions

are 500 °C and 1000 atm. Special pressure vessels called autoclaves are used in this technique.

(3) As mentioned in the last section, a large crystal can be grown by introducing a seed crystal into a high-temperature solution. This technique may be viewed as an extension of flux growth, or as a distinct method called top-seeded solution growth (TSSG), especially when an apparatus similar to that employed in the Czochralski method is used. Many commercial crystals of ferroelectric and non-linear optical materials are grown by this method.

(4) The travelling solvent floating zone (TSFZ) method uses the same appratus and setup as the floating zone method. Instead of moving up a melt of the same composition as the crystal, the TSFZ method carefully moves up a solution (usually solute plus self-flux) in the molten zone, allowing incongruently melting compounds to be grown. The best crystals of $La_{2-x}Sr_xCuO_4$ and several other transition-metal oxides are grown by this technique.

(5) In the electro-crystallization method, crystal growth occurs as a result of electrolysis in a high-temperature solution. Typically, a platinum wire is dipped into the solution to be used as the cathode, while a platinum crucible is used as the anode. Crystals grow at the cathode as an electric voltage or current is applied to the system. This technique can be used to grow single crystals of transition-metal compounds with unusual valency.

(6) In the high-pressure method, an environment of both high temperature (>1000 °C) and high pressure (>10^4 atm) is used to grow single crystals that are metastable under normal conditions. High pressure is used to convert one substance into another in some cases (such as graphite to diamond), and to drive chemical reactions that are prohibitive under normal pressure in other cases. Various novel compounds of interest to solid-state physicists have been grown under high pressure.

1.3.3 Vapor Growth Techniques

Some materials with high vapor pressures can turn into vapor without melting, a process known as sublimation. If the material is sealed in a tube (often made of silica glass) and placed in a horizontal tube furnace with an appropriate temperature gradient, crystals are deposited at different locations in the sealed tube. This technique of crystal growth is called the physical vapor transport method.

If direct sublimation is not feasible, a volatile solid or gas, called a transport agent, is added to the tube. At high temperatures, the transport agent will react with the material to form volatile intermediates at one end of the tube, which are then reacted back to the original constituents at the other end of the tube. The source material is often located at the hot zone and the crystals are deposited at the cold zone, but the reverse is also possible depending on the thermodynamics. This technique is called the chemical vapor transport method and is often used to obtain compounds containing volatile elements (such as S, Se, P, and As). In industry, various techniques of chemical vapor transport are used to obtain thin films of electronic materials.

1.3.4 Comparison of Different Methods

For some compounds, single crystals can be grown by more than one method (see Fig. 1.7), allowing the crystal grower to choose among different possibilities. In most cases, the largest crystals are grown from the melt. If crystals much larger than 1 cm^3 are needed, the melt techniques are often the best choice. (The TSSG and TSFZ methods can also be used to grow large crystals.) The melt techniques also produce crystals in the shortest amount of time, where crystals usually grow at a rate of 1 mm per hour or faster. This is to be compared with the typical rate of 1 mm per day or slower for solution growth, with even slower rates for vapor growth.

As for the quality of crystals, there is no simple answer as to which method produces the best crystals. This is because the method that produces the best crystals for one compound may not produce the best crystals for another compound, even if the two compounds are closely related. Moreover, the quality of crystals often depends on the expertise of the crystal grower and on the equipment; these are

Fig. 1.7 Ruby crystals of various origins. From left to right: A natural hexagonal ruby and several rubies embedded in marble rocks; a rod cut from a Czochralski-grown crystal for use in a laser; five flux-grown crystals (bottom); a hydrothermally grown crystal with a cobbled surface (top); two Verneuil-grown boules, where one of them has been split to relieve internal stress; three faceted pieces for jewelry, originally from Verneuil-grown boules. The length of the laser rod is 72 mm

especially crucial in the case of the Czochralski and TSSG methods. The question of crystal quality is also tricky because there are different types of imperfections found in crystals. In terms of impurities or chemical imperfections, the melt techniques that do not use a crucible (such as the floating zone and Verneuil methods) can produce crystals with minimum contamination. However, these techniques involve crystal growth under steep temperature gradients, which could lead to crystals with significant physical imperfections such as strains and dislocations.

Ultimately, the choice of growth method is often limited by the availability of equipment and expertise. In general, the flux method, aqueous solution method, Bridgman method, and physical and chemical vaport transport methods require the least amount of training on the part of the crystal grower. These methods also require the least expensive equipment, usually within the means of research groups that do not specialize in crystal growth. Because the flux method is arguably the most versatile of all, it is highly recommended for anyone who wants to start growing single crystals for physical research.

1.4 Literature on Flux Growth

The use of flux growth to obtain single crystals for solid-state research progressed rapidly in the 1960s and 1970s. Accordingly, many good resources come from this period. Probably the most well-known reference on flux growth is *Crystal Growth from High-Temperature Solutions* by Elwell and Scheel [16]. This book provides comprehensive coverage of all aspects of flux growth, with a list of all compounds grown by the flux method prior to 1975. Currently, a PDF version of this book can be downloaded, free of charge, from Scheel's website.

There are a number of review articles and book chapters on flux growth. Starting with Laudise [17], these include White [18], Schroeder and Linares [19], Laudise [20], Chase [21], Anthony and Collongues [22], Brice [23], Elwell [24], Wanklyn [25], Laudise [26], Elwell [27], Scheel [28], Wanklyn [29], Tolksdorf [30], Brice [31], Giess [32], Elwell [33], Pollert et al. [34], and Tolksdorf [35]. Although many of them cover similar materials, a particularly detailed discussion on growth defects is given by Chase [21], and Wanklyn [25] provides various tips on practical aspects. Recently, a review paper focusing on the flux growth of quaternary and higher order oxides has been published [36].

The above references mostly cover the growth of oxides, which has been the traditional subject of the flux method. The use of metallic flux to grow traditional semiconductors and related compounds was reviewed in the past [37–40]. More recently, there has been increasing interest in the use of metallic flux to grow novel intermetallic compounds for physical research. This area of flux growth is covered in the following review articles: Fisk and Remeika [41], Canfield and Fisk [42], Canfield and Fisher [43], Kanatzidis et al. [44], Thomas et al. [45], Canfield [46], and Phelan et al. [47].

A significant fraction of current work is reported in the *Journal of Crystal Growth*, which also publishes the reports of the conferences held by the International Conferences on Crystal Growth (ICCG). Review papers are published in *Progress in Crystal Growth and Characterization of Materials* (formally *Progress in Crystal Growth and Characterization*). Other journals on crystal growth, such as *Crystal Growth & Design*, *Crystal Research and Technology* (formally *Kristall und Technik*), and *CrystEngComm* also publish original works on flux growth. Some journals such as *Materials Research Bulletin*, *Philosophical Magazine*, *Journal of Materials Science*, and *Japanese Journal of Applied Physics* often contain papers with good descriptions of flux growth. Recently, *Philosophical Magazine* published a special issue (Vol. 92, Issue 19–21, 2012) on the design, discovery, and growth of novel materials, which contains many topical papers on flux growth. Of course, physics journals such as *Physical Review Letters* and *Physical Review B* publish many papers on the physical properties of flux-grown crystals, but these papers may not disclose the growth conditions in any detail. Chemistry journals such as *Journal of Solid State Chemistry* and *Chemistry of Materials* are good places to find information on new compounds.

References

1. M. Tachibana, H. Kawaji, T. Atake, Phys. Rev. B **70**, 064103 (2004)
2. T. Feder, Phys. Today **60**(8), 26–28 (2007)
3. P.C. Canfield, Nat. Phys. **4**, 167–169 (2008)
4. Y. Xia, D. Qian, D. Hsieh, L. Wray, A. Pal, H. Lin, A. Bansil, D. Grauer, Y.S. Hor, R.J. Cava, M.Z. Hasan, Nat. Phys. **5**, 398–402 (2009)
5. M.K. Wu, J.R. Ashburn, C.J. Torng, P.H. Hor, R.L. Meng, L. Gao, Z.J. Huang, Y.Q. Wang, C.W. Chu, Phys. Rev. Lett. **58**, 908–910 (1987)
6. R.J. Cava, B. Batlogg, R.B. van Dover, D.W. Murphy, S. Sunshine, T. Siegrist, J.P. Remeika, E.A. Rietman, S. Zahurak, G.P. Espinosa, Phys. Rev. Lett. **58**, 1676–1679 (1987)
7. H. Kikuchi, Y. Fujii, M. Chiba, S. Mitsudo, T. Idehara, T. Tonegawa, K. Okamoto, T. Sakai, T. Kuwai, H. Ohta, Phys. Rev. Lett. **94**, 227201 (2005)
8. P. Gehring, H.M. Benia, Y. Weng, R. Dinnebier, C.R. Ast, M. Burghard, K. Kern, Nano Lett. **13**, 1179–1184 (2013)
9. T.H. Maiman, Nature **187**, 493–494 (1960)
10. R.C. Linares, J. Phys. Chem. Solids **26**, 1817–1820 (1965)
11. D. Aoki, Y. Haga, T.D. Matsuda, N. Tateiwa, S. Ikeda, Y. Homma, H. Sakai, Y. Shiokawa, E. Yamamoto, A. Nakamura, R. Settai, Y. Onuki, J. Phys. Soc. Jpn. **76**, 063701 (2007)
12. R. Liang, D.A. Bonn, W.N. Hardy, Philos. Mag. **92**, 2563–2581 (2012)
13. Y. Hidaka, M. Suzuki, Prog. Cryst. Growth Charact. Mater. **23**, 179–243 (1991)
14. N. Ghosh, S. Elizabeth, H.L. Bhat, U.K. Rößler, K. Nenkov, S. Rößler, K. Dörr, K.-H. Müller, Phys. Rev. B **70**, 184436 (2004)
15. T. Yankova, D. Hüvonen, S. Mühlbauer, D. Schmidiger, E. Wulf, S. Zhao, A. Zheldev, T. Hong, V.O. Garlea, R. Custelcean, G. Ehlers, Philos. Mag. **92**, 2629–2647 (2012)
16. D. Elwell, H.J. Scheel, *Crystal Growth from High-Temperature Solutions* (Academic Press, London, 1975)
17. R.A. Laudise, in *The Art and Science of Growing Crystals*, ed. by J. J. Gilman (Wiley, New York, 1963), pp. 252–273

18. E.A.D. White, in *Technique of Inorganic Chemistry*, vol. 4, ed. by H.B. Jonassen, A. Weissberger (Wiley, New York, 1965), pp. 31–64
19. J.B. Schroeder, R.C. Linares, in *Progress in Ceramic Science*, vol. 4, ed. by J.E. Burke (Academic Press, London, 1966), pp. 196–216
20. R.A. Laudise, *The Growing of Single Crystals* (Prentice Hall, Engelwood Cliffs, N. J., 1970)
21. A.B. Chase, in *Preparation and Properties of Solid State Materials*, vol. 1, ed. by R.A. Lefever (Dekker, New York, 1971), pp. 183–264
22. A.M. Anthony, R. Collongues, in *Preparative Methods in Solid State Chemistry*, ed. by P. Hagenmuller (Academic Press, New York, 1972), pp. 147–249
23. J.C. Brice, *The Growth of Crystals from Liquids* (North-Holland, Amsterdam, 1973)
24. D. Elwell, in *Crystal Growth*, ed. by B.R. Pamplin (Pergamon, Oxford, 1975), pp. 185–216
25. B. Wanklyn, in *Crystal Growth*, ed. by B.R. Pamplin (Pergamon, Oxford, 1975), pp. 217–288
26. R.A. Laudise, in *Treatise on Solid State Chemistry*, vol. 5, ed. by N.B. Hannay (Plenum Press, New York, 1975), pp. 407–461
27. D. Elwell, in *Crystal Growth, Second edition*, ed. by B.R. Pamplin (Pergamon, Oxford, 1980), pp. 463–483
28. H.J. Scheel, Prog. Cryst. Growth Charact. **5**, 277–290 (1982)
29. B. Wanklyn, J. Cryst. Growth **65**, 533–540 (1983)
30. W. Tolksdorf, in *Crystal Growth of Electronic Materials*, ed. by E. Kaldis (Elsevier, Amsterdam, 1985), pp. 175–182
31. J.C. Brice, *Crystal Growth Processes* (Blackie, Glasgow, 1986)
32. E.A. Giess, in *Advanced Crystal Growth*, ed. by P.M. Dryburgh, B. Cockayne, K.G. Barraclough (Prentice Hall, London, 1987), pp. 245–265
33. D. Elwell, in *Crystal Growth in Science and Technology*, ed. by H. Ahrend, J. Hulliger (Plenum Press, New York, 1989), pp. 133–142
34. E. Pollert, M. Nevřiva, S. Durčok, Prog. Cryst. Growth Charact. Mater. **22**, 143–182 (1991)
35. W. Tolksdorf, in *Handbook of Crystal Growth*, vol. 2, ed. by D.T.J. Hurle (Elsevier, Amsterdam, 1994), pp. 563–611
36. D.E. Bugaris, H.-C. zur Loye, Angew. *Chem. Int. Ed.* **51**, 3780–3811 (2012)
37. N.P. Luzhnaya, J. Cryst. Growth **3–4**, 97–107 (1968)
38. R.H. Deitch, in *Crystal Growth*, ed. by B.R. Pamplin (Pergamon, Oxford, 1975), pp. 428–496
39. V.N. Gurin, M.M. Korsukova, Prog. Cryst. Growth Charact. **6**, 59–101 (1983)
40. T. Lundström, J. Less-Common Metals **100**, 215–228 (1984)
41. Z. Fisk, J.P. Remeika, in *Handbook on the Physics and Chemistry of Rare Earths*, vol. 12, ed. by K.A. Gschneider, Jr., L. Eyring (Elsevier, Amsterdam, 1989), pp. 53–70
42. P.C. Canfield, Z. Fisk, Philos. Mag. B **65**, 1117–1123 (1992)
43. P.C. Canfield, I.R. Fisher, J. Cryst. Growth **225**, 155–161 (2001)
44. M.G. Kanatzidis, R. Pöttgen, W. Jeitschko, Angew. *Chem. Int. Ed.* **44**, 6996–7023 (2005)
45. E.L. Thomas, J.N. Millican, E.K. Okudzeto, J.Y. Chan, Comments Inorg. Chem. **27**, 1–39 (2006)
46. P.C. Canfield, in *Properties and Applications of Complex Intermetallics*, ed. by E. Belin-Ferré (World Scientific, Singapore, 2010), pp. 93–111
47. W.A. Phelan, M.C. Menard, M.J. Kandas, G.T. McCandless, B.L. Drake, J.Y. Chan, Chem. Mater. **24**, 409–420 (2012)
48. C. Petrovic, P.G. Pagliuso, M.F. Hundley, R. Movshovich, J.L. Sarrao, J.D. Thompson, Z. Fisk, P. Monthoux, J. Phys.: Condens. Matter **13**, L337–L342 (2001)
49. R.E. Baumbach, Z. Fisk, F. Ronning, R. Movshovich, J.D. Thompson, E.D. Bauer, Philos. Mag. **94**, 3663–3671 (2014)
50. M.D. Lumsden, B.D. Gaulin, H. Dabkowska, M.L. Plumer, Phys. Rev. Lett. **76**, 4919–4922 (1996)
51. B.M. Wanklyn, A. Maqsood, J. Mater. Sci. **14**, 1975–1981 (1979)
52. A. Kania, A. Słodczyk, Z. Ujma, J. Cryst. Growth **289**, 134–139 (2006)
53. B.M. Wanklyn, J. Mater. Sci. **7**, 813–821 (1972)

54. R. Palai, R.S. Katiyar, H. Schmid, P. Tissot, S.J. Clark, J. Robertson, S.A.T. Redfern, G. Catalan, J.F. Scott, Phys. Rev. B **77**, 014110 (2008)
55. D.P. Chen, C.T. Lin, Supercond. Sci. Technol. **27**, 103002 (2014)
56. I.R. Fisher, M.C. Shapiro, J.G. Analytis, Philos. Mag. **92**, 2401–2435 (2012)
57. V.N. Gurin, M.M. Korsukova, S.P. Nikanorov, I.A. Smirnov, N.N. Stepanov, S.G. Shul'man, J. Less-Common Met. **67**, 115–123 (1979)

Chapter 2
Mechanisms of Crystal Growth from Fluxed Solutions

It was stated in the previous chapter that flux growth can produce high-quality crystals. Although this is an encouraging statement, it probably demands further explanation: What exactly is meant by high-quality crystals, and under what conditions do high-quality crystals grow? These are the questions that can be answered with some understanding of the crystal growth mechanisms, the topic of this chapter.

Although there are various references on the theories of crystal growth, most are written in such a way as to make very heavy going of the subject. This chapter, by contrast, attempts to make the subject approachable for anyone who wishes to obtain basic ideas for growing high-quality crystals. For further details, the following books are especially recommended: *Crystals: Growth, Morphology and Perfection* by Sunagawa [1] and *Crystal Growth from High-Temperature Solutions* by Elwell and Scheel [2]. The figures used in these books were helpful in preparing some of the illustrations presented in this chapter.

2.1 Crystal Morphology

By definition, atoms in a crystal are arranged in an orderly, repetitive array. This internal regularity often shows itself on the outside of the crystal, as flat faces meeting at sharp edges and pointed corners. However, the appearance of a crystal is also affected by the manner in which it was grown—some crystals grow into perfect polyhedra, while others have irregular shapes. For this reason, crystal morphology (crystal shape) is closely related to the study of crystal growth, and this is why we should take a look at this topic before proceeding directly to the growth mechanisms. Crystal morphology is also important for solid-state research, because the shape often limits the usefulness of the crystal as a sample specimen.

M. Tachibana, *Beginner's Guide to Flux Crystal Growth*, NIMS Monographs,
DOI 10.1007/978-4-431-56587-1_2

Many compounds of interest to solid-state physicists have complex atomic arrangements of crystal structures. However, it is not often necessary to think about every atom in the crystal structure—what we can do instead is to replace the repeating unit by a single point. Therefore, the point may correspond to a single atom in simple structures, or to a group of many atoms in complicated structures. The resulting three-dimensional array made up of these points is called a lattice, from which we can choose a unit cell defined by three axes a, b, and c and three angles α, β, and γ. On the basis of these six parameters, seven crystal systems can be identified: cubic, tetragonal, orthorhombic, hexagonal, rhombohedral, monoclinic, and triclinic. There are 14 different ways of arranging points in space, giving rise to the 14 Bravais lattices (Fig. 2.1). Furthermore, there are 32 independent ways of arranging objects about a point, which make up the 32 point groups, and 230 possible arrangements of objects in space, which are called the 230 space groups.

The 7 crystal systems, 14 Bravais lattices, 32 point groups, and 230 space groups can be used to describe the internal symmetry of crystals. The concepts of symmetry are also useful in understanding the crystal shapes, especially when the shapes are described using the language of crystal form. A crystal form, as defined in crystallography, is a set of identical faces that are related by symmetry. This sounds very abstract, so let us look at real examples.

In the top row of Fig. 2.2, three crystals of different shapes are shown. The left crystal is perovskite $La_{1-x}Pr_xAlO_3$, which has the shape of a cube with six identical square faces. This description of a cube satisfies the definition of a crystal form; therefore, the cube is one type of crystal form. Each face of the cube has a form symbol of {100}, because it is perpendicular to one of the three cubic crystallographical axes and does not intersect with the other two axes. The middle crystal is spinel $MgAl_2O_4$, which has the shape of an octahedron. The octahedron is composed of eight triangular faces, so it is also a type of crystal form. In this case, the faces have a form symbol of {111}, as each face intersects with three cubic axes at the same distance from their origin at the center of the octahedron. Finally, the crystal on the right side is garnet $Y_{3-x}Ce_xAl_5O_{12}$, which appears to have twelve rhombic faces to form a dodecahedron. The dodecahedron is yet another example of crystal form, with the rhombic face having the form symbol of {110}. The cube, octahedron, and dodecahedron are thus three examples of crystal form, each belonging to the cubic system. (In other crystal systems, a crystal form often does not make a complete body on its own. For example, the four side faces of a tetragonal body, called a tetragonal prism, require another form on the top and bottom to make a closed body. Such a form is called an open form, as opposed to the closed forms of the cubic examples. The topic of crystal forms is fully described in mineralogy textbooks.)

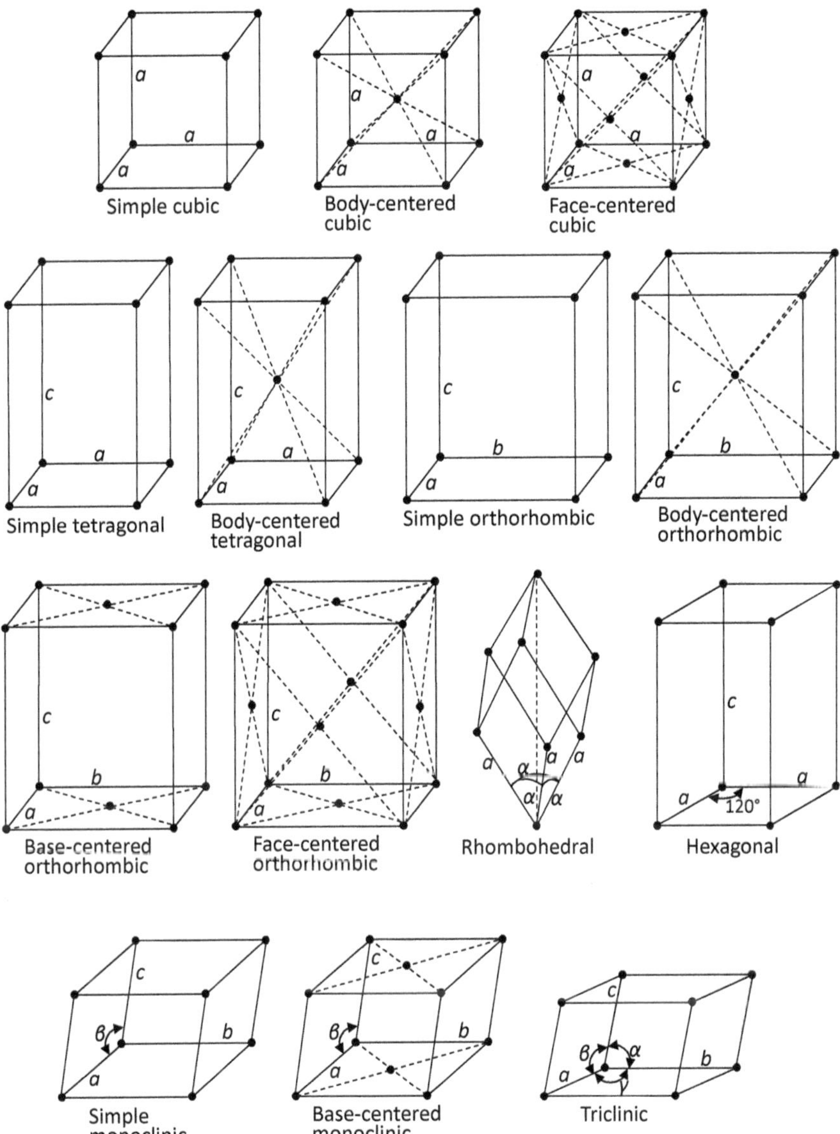

Fig. 2.1 Unit cells of the 14 Bravais lattices, shown with the constraints on the length of edges and angles between edges

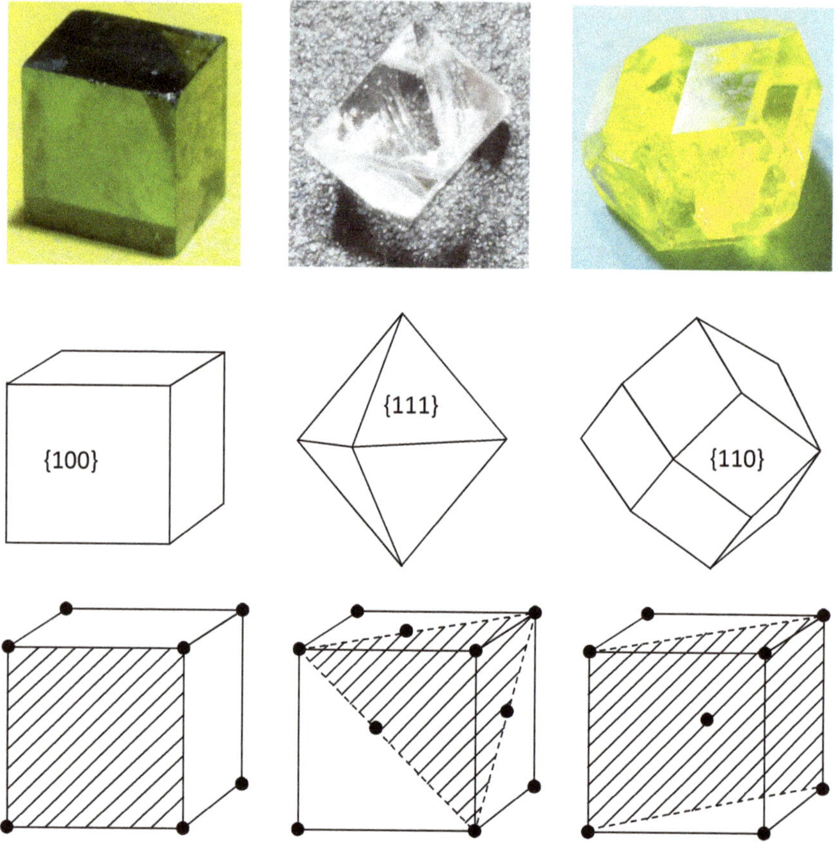

Fig. 2.2 Top: Single crystals of perovskite $La_{1-x}Pr_xAlO_3$ (left), spinel $MgAl_2O_4$ (middle), and garnet $Y_{3-x}Ce_xAl_5O_{12}$ (right). Center: Cube, octahedron, and dodecahedron crystal forms. Bottom: Simple cubic, face-centered cubic, and body-centered cubic lattices

In addition to the particular crystals shown in Fig. 2.2, other crystals of perovskites often have the form of a cube, those of spinels have the octahedron form, and those of garnets have the dodecahedron form. Perovskites, spinels, and garnets each refer to a group of compounds with the same basically cubic crystal structure as the mineral of the same name.[1]

What, then, determines the form of these crystals? A hint comes from the fact that while these compounds all belong to the cubic system, they have different Bravais lattices: the perovskite structure has a simple (primitive) cubic lattice, the spinel structure has a face-centered cubic lattice, and the garnet structure has a

[1]The word "basically" is added because many perovskite compounds, including $La_{1-x}Pr_xAlO_3$, undergo a subtle transition from the cubic perovskite structure on cooling. However, this has no effect on the crystal form.

body-centered cubic lattice. These lattices can be found in Fig. 2.1, and are reproduced at the bottom of Fig. 2.2. Here, we remind ourselves that we are not looking at the actual atomic arrangements, which are much more complex (for example, the unit cell of garnets contains 160 atoms).

According to the Bravais principle, crystal faces are most likely to come from the most densely populated lattice planes. As shown in the bottom images of Fig. 2.2, the simple cubic lattice has a lattice point at each of the eight corners; accordingly, the {100} plane has the highest density of lattice points, and this agrees with the cube form of the $La_{1-x}Pr_xAlO_3$ crystal. We can see from Fig. 2.2 that similar arguments apply to the octahedron coming from the face-centered cubic lattice, as well as the dodecahedron from the body-centered cubic lattice. It makes sense that densely populated lattice planes become the crystal faces, as these planes should be thermodynamically and mechanically more stable than sparsely populated lattice planes.

Unfortunately, the simple picture presented in terms of only the Bravais lattice is incomplete. For example, garnets often have {210} faces, either by themselves or in combination with the {110} faces discussed above. The {210} faces come from another type of crystal form, called the trapezohedron, which is a regular polyhedron with 24 sets of four-sided faces.[2] To explain such observations, it is necessary to use the full symmetry considerations of garnet's space group, rather than just those of the Bravais lattice [3]. In other cases, the direction of chemical bonds becomes important, and this is considered in the periodic bond chain (PBC) model [4].

As the case of garnet has shown, faces of more than one form can appear on a crystal. For example, Fig. 2.3a shows various shapes that are made of the cube, octahedron, and dodecahedron, and Fig. 2.3b shows single crystals of $Pb_2Ru_2O_{6.5}$ with a shape that is intermediate between the cube and octahedron. In crystal morphology, the term "habit" is used for the characteristic shape of a crystal, and we say that each object in Fig. 2.3a has a different habit. It is also possible for different habits to arise from a single form, if the faces develop to different sizes during crystal growth (*see* Fig. 2.4). In each case, the change in habit modifies the shape of crystal faces, but retains the characteristic angles between corresponding faces. The crystal habit can be affected by changes in various growth conditions, such as growth temperature (Fig. 2.5a) [5], flux composition (Fig. 2.5b) [6], impurities or additives present, and supersaturation. The concept of supersaturation will be discussed in the next section.

Surprisingly enough, it is the fast-growing faces that disappear and the slowest growing ones that determine the final crystal habit—*see* Fig. 2.6a. This idea can also be understood by picturing an object that is growing from many small simple

---

[2]Indeed, small {210} faces can be seen on the $Y_{3-x}Ce_xAl_5O_{12}$ crystal of Fig. 2.2, which is why the crystal does not quite have the shape of a perfect dodecahedron.

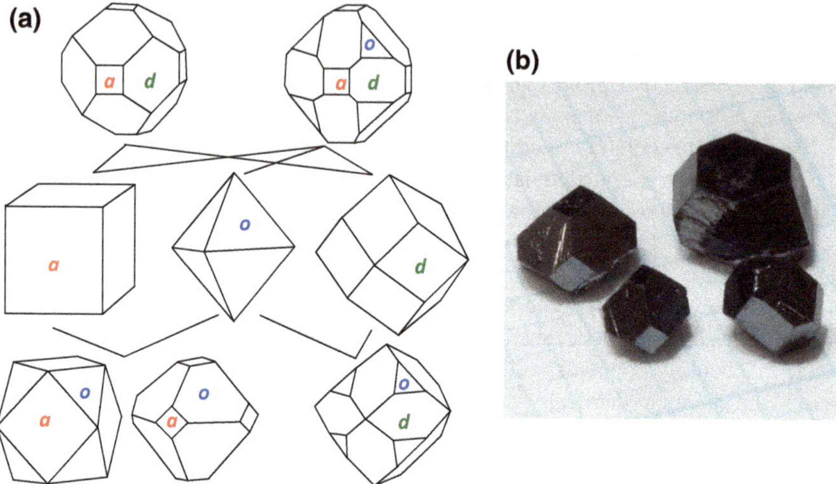

Fig. 2.3 **a** Various habits originating from the cube, octahedron, and dodecahedron crystal forms. The form symbols $a\{100\}$, $o\{111\}$, and $d\{110\}$ are shown, where the letter symbols are frequently used in mineralogical works. **b** Single crystals of $Pb_2Ru_2O_{6.5}$ showing a combination of {100} and {111} cubic faces

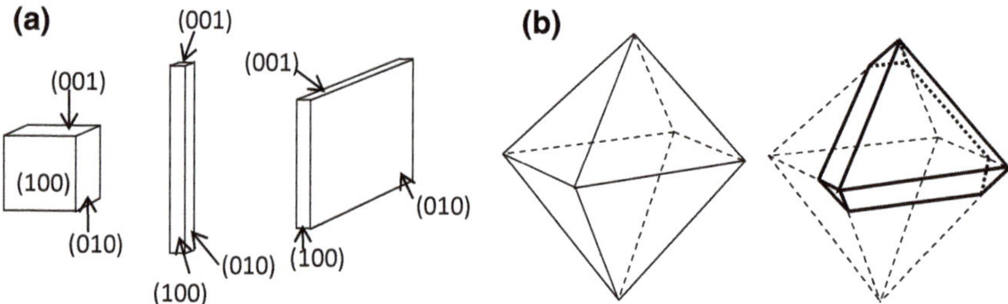

Fig. 2.4 **a** Different habits originating from the cube form. (100), (010), and (001) are the specific faces of the {100} form. **b** Tabular crystal originating from the octahedral form

cubes (Fig. 2.6b): At the corner, the diagonal face is always rough, providing many bonding surfaces for the small cubes. Consequently, the growth rate of this rough face is much faster than that of the smooth faces, and soon the entire crystal consists of the six flat faces of a cube.

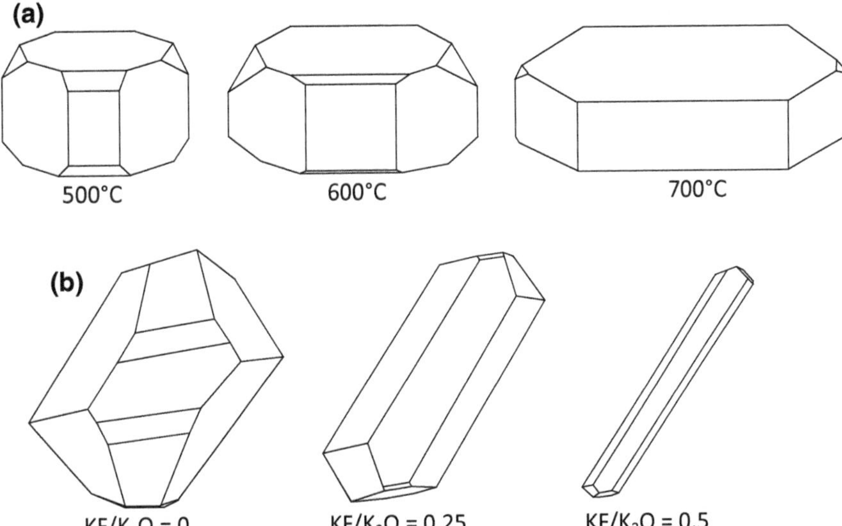

Fig. 2.5 **a** Morphological change of NdP_5O_{14} crystals (simple monoclinic structure) owing to a difference in the growth temperature (after [5]). **b** Morphological change of $NdAl_3(BO_3)_4$ crystals (base-centered monoclinic structure) owing to a difference in the flux composition (after [6])

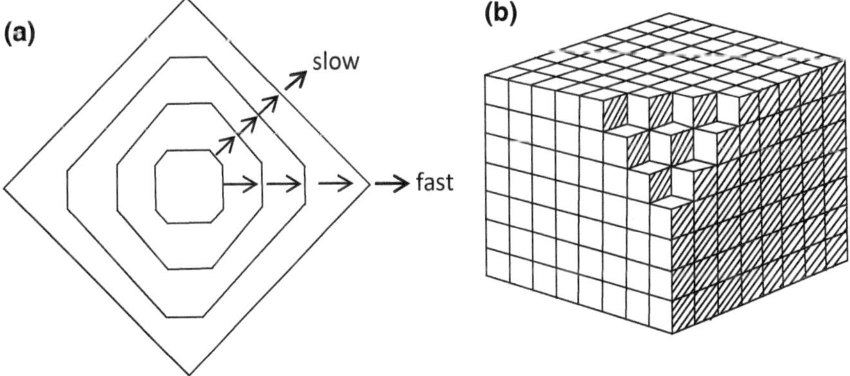

Fig. 2.6 **a** Effect of growth rate on crystal faces. Slow-growing faces survive and rapidly growing faces disappear. **b** A model of a crystal composed of small cubes. The diagonal face is always rough, and incoming growth units of cubes can strongly adhere to this face

2.2 Mechanisms of Flux Crystal Growth

Having looked at the basic ideas of crystal morphology, let us now focus our attention on how crystals grow. In crystal growth from solution, there are three major steps: (1) the attainment of supersaturation, (2) the formation of nuclei of the crystalline phase, and (3) subsequent crystal growth on the nuclei. We look at each of these steps in this section.

2.2.1 Solubility and Supersaturation

Probably the most important concept in flux growth is supersaturation. Without supersaturation, there is no driving force for crystals to appear in a solution, or to grow larger once they appear in the solution. Being a thermodynamic quantity, supersaturation is related to the decrease in free energy resulting from the growth of crystals. However, there is no need to go into thermodynamics for practical purposes, because supersaturation can be simply defined using a solubility curve.

The solubility curve, as shown in Fig. 2.7, is a plot of the maximum amount of solute that can be dissolved into the solution at each temperature. The region below the solubility curve represents an unsaturated solution; here, the solution can dissolve more solute. On the other hand, the region above the solubility curve indicates a supersaturated solution—the solution contains more than the equilibrium concentration of solute, and it is therefore thermodynamically unstable. It is only at the solubility curve, corresponding to the saturated solution, where a solute crystal in contact with the solution neither dissolves nor grows. The solubility curve in Fig. 2.7 has a positive slope, indicating that more solute can be dissolved at higher temperatures. This is what happens in most solutions.

When a solution is unsaturated, it is thermodynamically stable and no crystal will ever form. However, if the temperature of such a solution is lowered to the

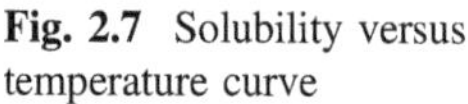

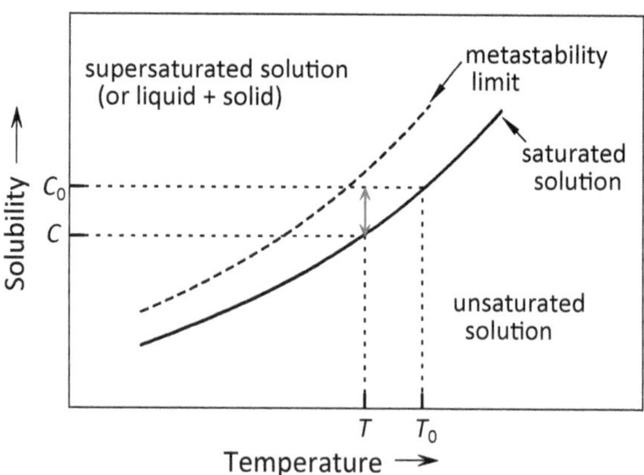

Fig. 2.7 Solubility versus temperature curve

region of supersaturation, there is now more solute in the solution than it can actually handle. Therefore, the excess solute must somehow crystallize out of the solution.

However, crystals do not appear immediately after the solution becomes supersaturated. This is due to the presence of an energy barrier, which we will see in a short while, but here we point out that the solution is now in a metastable region. Such a metastable region is shown in Fig. 2.7 as the area surrounded by the solubility curve and the dashed curve.

We can define supersaturation as follows. If we start with a saturated solution at (T_0, C_0) and lower the temperature to T, the equilibrium concentration of solute decreases to C. This difference in solubility ($C_0 - C$) is the excess solute in the solution at T, so the supersaturation σ at this temperature can be defined as $\sigma = (C_0 - C)/C$. Evidently, a large σ means that there is a strong driving force for crystallization. It is also evident that the temperature must be lowered during crystal growth to maintain a finite σ, as crystallization removes the excess solute in the solution and lowers the σ. (σ can also be increased by evaporating the flux.)

Although we are looking at supersaturation in the case of crystal growth from solution, it is useful to mention that crystal growth from the vapor occurs when the vapor is supersaturated, and crystal growth from the melt occurs when the melt is supercooled. Supersaturation and supercooling are therefore the driving forces of crystallization in these cases.

2.2.2 *Nucleation*

As we have stated earlier, a supersaturated solution is metastable. This means that more stable states can be achieved by forming crystals, which reduces the amount of excess solute dissolved in the solution. However, for crystals to start appearing in the solution, it is first necessary for solute particles to come together and form stable nuclei. This process is called nucleation.

The process of nucleation is similar to the formation of droplets in a supersaturated vapor (*see* Fig. 2.8). As vapor molecules move about randomly, fluctuations within the supersaturated vapor give rise to small clusters of molecules. The chance that such clusters will grow to form stable droplets depends on their total free energy. On one hand, the bulk free energy of a droplet is always lower than that of vapor under supersaturation. This energy is proportional to the volume of the droplet, or r^3, where r is the radius of the droplet. On the other hand, a droplet has additional interfacial surface energy, because its outermost molecules are under-bonded and highly strained. This energy is proportional to the surface area of the droplet, or r^2.

When the radius dependences of both bulk free energy and surface energy are considered, we can find that the total free energy of a droplet goes through a maximum barrier at r^*, which is called the critical radius. This implies that a droplet with a radius smaller than r^* becomes more stable by reducing its size, meaning

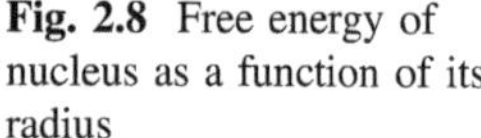

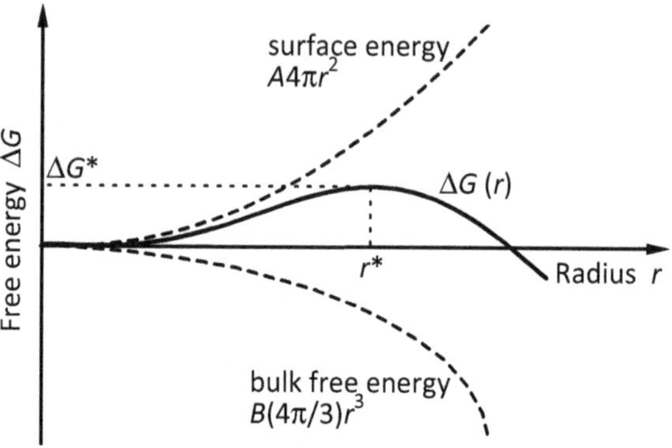

Fig. 2.8 Free energy of nucleus as a function of its radius

that it will evaporate and disappear. On the other hand, once a droplet reaches r^*, it is not likely to disappear because it can reduce its free energy by increasing its size. An increase in supersaturation has the effect of reducing of r^*.

A similar situation is found for nucleation in a supersaturated solution. Because solute particles (atoms, ions, or molecules) are in random motion, there is always a chance that some particles will come into contact with each other and form a small cluster. Most of these clusters soon dissipate back into the solution. However, when the supersaturation is high, gathering of more particles is encouraged and some clusters reach r^*, becoming stable nuclei. This process is often referred to as homogeneous nucleation, because no other substances are involved. Various studies have shown that the critical size of the nucleus is of the order of 100 atoms [2].

In real experiments, homogeneous nucleation does not occur frequently. This is because the energy barrier for nucleation is reduced if it occurs on the surface of other substances—on the crucible wall, the surface of the solution, and even some dust particles in the solution. This is called heterogeneous nucleation, and much evidence suggests that this is usually the case in flux growth. (In Fig. 1.6, ruby crystals are grown on the crucible wall.) Furthermore, the energy barrier for nucleation can be completely diminished if a seed crystal of the same material is introduced into the solution. Because crystal growth can start immediately on the seed, it can prevent spurious nucleation from occurring elsewhere in the solution.

In general, nucleation requires much higher supersaturation than the subsequent crystal growth on the nucleus. This is because nucleation is a process of creating initial order from random solute particles, whereas subsequent crystal growth is a process of attaching additional atoms onto an already ordered array. If supersaturation is kept low after the initial nucleation, crystal growth can take place without additional nuclei forming in the solution. This can be done in experiments by slowly cooling the solution. In many cases, the real challenge is to reduce the number of initial nuclei forming in the solution, so that a small number of large crystals, rather than a large number of small crystals, result at the end of crystal growth.

2.2.3 *Layer-by-Layer Growth*

When a stable nucleus is formed, it is expected to be nearly spherical in shape. It can grow larger through attachment of more solute particles, which reduces the free energy (*see* Fig. 2.8). As this process takes solute particles away from the solution, a thin layer is formed around the growing crystal where the solute is more diluted (less supersaturated) than the bulk of the solution. This concentration difference drives solute particles in the solution to diffuse into the thin layer region and then to attach to the surface of the crystal.

In the past, this simple mechanism was once considered to continue throughout crystal growth. However, it soon became evident that this mechanism will only result in spherical crystals, in contrast to the polyhedral shapes and flat faces found in real crystals. Clearly, there was something missing in the picture. A clue to solving this problem is the presence of various terraced features on crystal faces (Fig. 2.9), which suggests that crystals somehow grow layer upon layer on flat faces.

To explain the flat faces of crystals, the layer-by-layer growth mechanism, on an atomically smooth surface (*see* Fig. 2.10), was introduced in the 1930s [1, 2]. In this mechanism, the solute particles reaching the crystal surface are not immediately incorporated into the crystal, but instead form growth units that are loosely adsorbed on the surface. These units can wander around the surface to find a suitable site for attachment, such as the kinks along the step. Kinks are the preferred site because the growth unit can bond with three or four units of the crystal, compared with just one unit on a flat surface. Some growth units are dissolved back into the solution before reaching a kink. The addition of more units along the kinks and steps causes the layer to spread laterally across the surface.

So far, we have not specified what the solute particles and growth units are made of. They can be atoms, ions, groups of ions, or molecules, depending on the nature of the solute and its interaction with the flux particles. Because a solute particle is bounded by flux particles in the solution, it sheds some of the flux particles upon

Fig. 2.9 Terraced features on the faces of $Y_3Fe_5O_{12}$ (left), $DyMn_2O_5$ (middle), and $Al_{2-x}Cr_xO_3$ (right) crystals

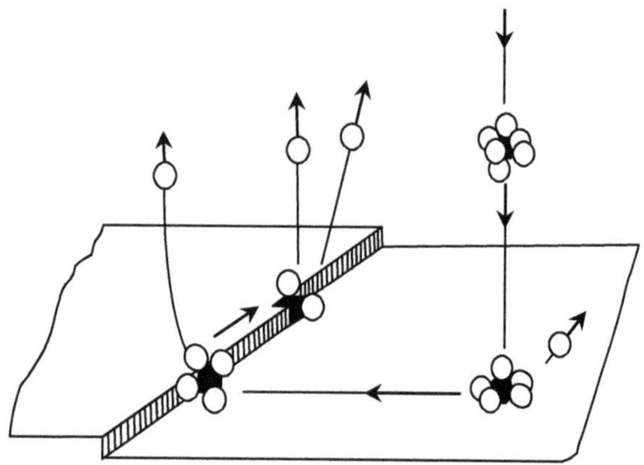

Fig. 2.10 Schematic of the layer-by-layer growth mechanism due to two-dimensional nucleation. The growth unit is surrounded by flux particles, which are fully desorbed when the growth unit becomes firmly attached at the kinked site along the step. Based on [2]

reaching the crystal surface. This process is called desolvation. As show in Fig. 2.10, there are additional steps of desolvation until the growth unit is firmly attached to the crystal.

2.2.4 *Spiral Growth*

The layer-by-layer growth mechanism can explain the flat faces of crystals. However, there is one problem with this mechanism: When the layer is completed, there is no easy place for the new growth unit to attach onto the crystal, and crystal growth cannot continue without starting a new layer. Detailed calculations show that supersaturation of well above 10% must be present for a new layer to nucleate on a completed surface. In reality, however, crystal growth can continue at much lower supersaturation, sometimes well below 1%.

To solve this problem, the spiral growth mechanism was introduced in 1949 [7, 8].[3] In this mechanism, a screw dislocation provides a permanent step on the crystal surface. (Screw dislocation, shown in Fig. 2.11a, is a defect that can be visualized by imagining a partial cut into a crystal, followed by a slight twist.) As depicted in Fig. 2.11b, the spiral growth takes place by the rotation of a step around the central point—the layer is never completed. Therefore, this mechanism allows the crystal to grow continuously at low supersaturation, since the nucleation of a new layer is not required.

The theory of spiral growth was soon verified by the observation of spirals on many crystal surfaces; two of the more recent examples are shown in Fig. 2.12 [9, 10]. There are many types of spirals. Both rounded and polygon-shaped spirals are found. The height of the step can be anywhere from one growth unit to hundreds of

[3]The ideas of layer-by-layer growth and spiral growth were initially introduced for crystal growth from vapors, but they also apply to crystal growth from solutions.

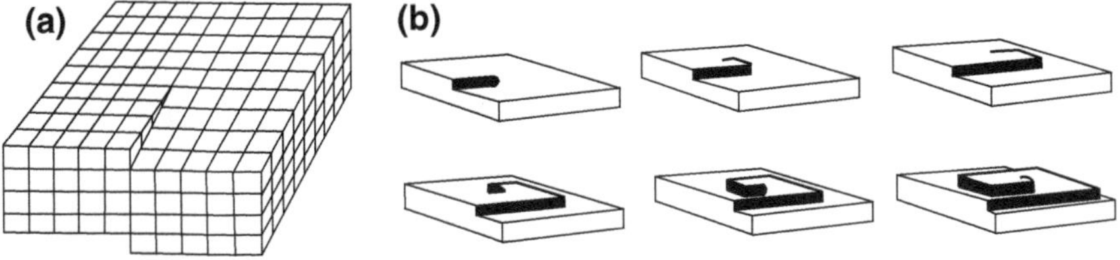

Fig. 2.11 **a** Screw dislocation. **b** Spiral growth around screw dislocation

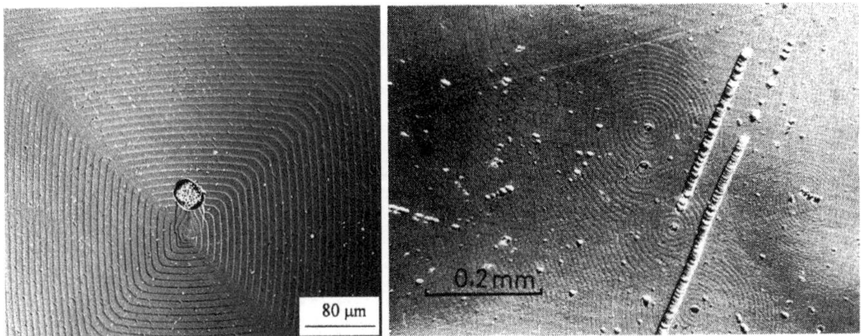

Fig. 2.12 Spiral growth steps on crystal faces of $YBa_2Cu_3O_7$ (left, reprinted with permission from [9]; copyright 2000 Elsevier) and $Sm_{0.55}Tb_{0.45}FeO_3$ (right, reprinted with permission from [10]; copyright 1972 The Japan Society of Applied Physics). Two rows of etch pits, which signal arrays of edge dislocations, can be seen on $Sm_{0.55}Tb_{0.45}FeO_3$

growth units—the latter occurs when the growing layers are bunched together by impurities or other factors. Because each crystal face often contains a number of screw dislocations, several spirals may appear on the same face. It should be emphasized that the spiral growth mechanism does not invalidate the basic idea of the layer-by-layer mechanism, because the process of a layer spreading across the crystal surface is the same. Rather, the spiral growth mechanism provides a source of steps that will not go away.

2.2.5 Hopper Growth and Dendritic Growth

Spiral growth takes place when the supersaturation is not sufficiently high to allow a new layer to nucleate on a flat crystal surface. Conversely, the layer-by-layer mechanism can take place if the supersaturation is sufficiently high to allow surface nucleation, and indeed, this mechanism can achieve higher growth rates than spiral growth under high supersaturation. One point to remember, however, is that because the corners and edges of a crystal overlook a greater volume of solution than the face center, the supersaturation at these regions is higher (this is known as

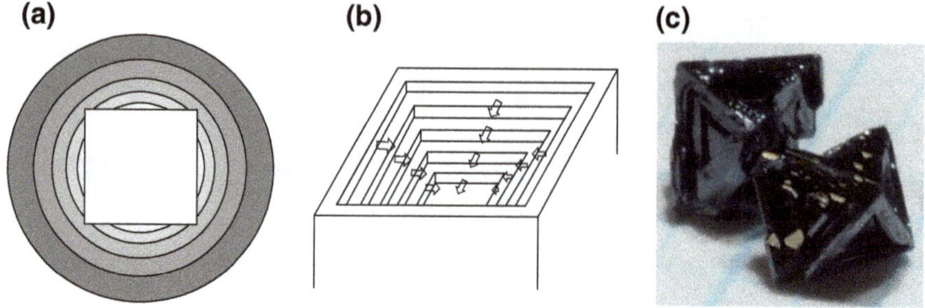

Fig. 2.13 **a** Contours of constant solute concentration (supersaturation) around a growing crystal. The darker/outer region has a higher solute concentration. **b** Hopper morphology on a cube face. **c** Octahedral hopper crystals of $Bi_2Ru_2O_7$

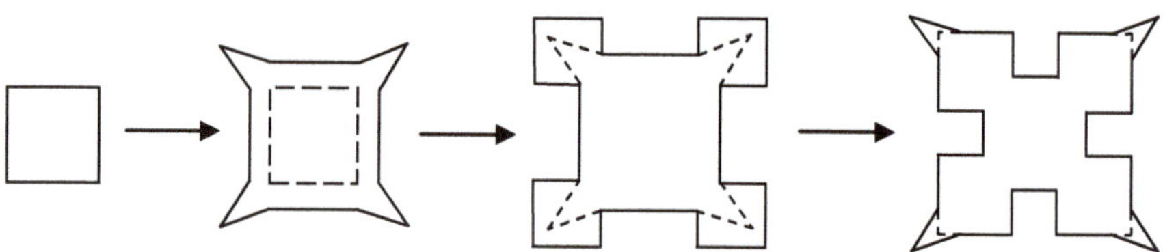

Fig. 2.14 Dendritic growth

the Berg effect; *see* Fig. 2.13a). Therefore, in the layer-by-layer mechanism, a new layer tends to start at the edges and corners of a crystal face, and then spreads toward the center region.

Flat crystal faces are maintained only when layer propagation occurs at a higher rate than layer nucleation. When supersaturation is high, additional layers can be nucleated before the underlying layers are completed. This situation is shown for the square face of a cube in Fig. 2.13b. The resulting crystals are called hopper or skeletal crystals; an example of octahedral hopper crystals is shown in Fig. 2.13c.

If supersaturation is higher still, new layers formed at the edges and corners of a crystal can protrude into the solution, rather than spread across the face (Fig. 2.14). This is called dendritic growth, and the resulting treelike crystals with many branches are called dendrites. Some examples of dendritic crystals are shown in Fig. 2.15.

In flux growth, dendritic growth often occurs at the beginning of crystal growth, when supersaturation is high. As the crystal growth continues and supersaturation drops, the spaces between dendrite arms are filled and flat faces are eventually produced. Spiral growth can then take place on such flat faces.

Fig. 2.15 Dendritic crystals of $Tb_2Ge_2O_7$ (left), $Na_{1/2}Bi_{1/2}TiO_3$ (middle, three crystals), and $Bi_4Ti_3O_{12}$ (right)

2.2.6 Summary of Growth Mechanisms

The relationships among spiral growth, hopper growth by the layer-by-layer mechanism, and dendritic growth are summarized [1] in Fig. 2.16. In general, crystals grow faster under higher supersaturation. When supersaturation is very low, surface nucleation is not possible and only spiral growth can take place. However, when supersaturation exceeds σ^*, nucleation of new layers occurs at the edges and corners of the crystal and hopper growth becomes dominant. Both spiral growth and hopper growth are characterized by the spreading of layers at the crystal surface. When supersaturation exceeds σ^{**}, dendritic growth becomes operative. Because dendritic growth occurs on atomically rough surfaces, the growth rate can be very high.

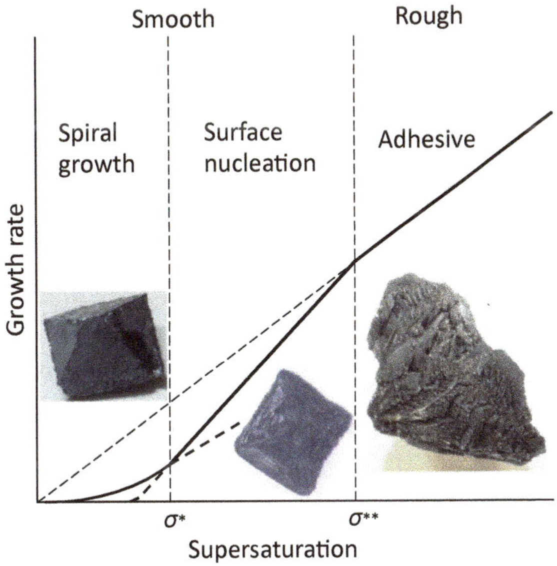

Fig. 2.16 Growth mechanisms as a function of supersaturation. The solid line is the growth rate in each regime of supersaturation. Crystals of Mn_3O_4 grown under each regime are also shown

The above discussion shows that crystals with flat, smooth faces are grown when supersaturation is kept low. Because such growth proceeds slowly, usually at a rate of about 1 mm per day or slower, the accidental formation of various defects can be minimized. The condition of low supersaturation can be achieved by the slow cooling of the solution, but real growth processes often involve interactions of different mechanisms and it is always difficult to predict the exact outcome of a given experiment. Factors such as the viscosity of the solution and temperature instabilities often influence the quality of crystals. Such factors will be discussed in later chapters.

2.3 Imperfections in Crystals

Various types of imperfection are introduced into growing crystals. Crystal imperfections include physical defects, chemical impurities, and various forms of inhomogeneities. In some cases, the most visible imperfection is twinning (Fig. 2.17). A twin is a composite of two or more crystals, each showing a definite crystallographic relationship to the others. In Fig. 2.17b, the twinning plane, where the two parts share atoms along a common plane, is shown as the area surrounded by dashed lines. Most twins originate during the nucleation stage of crystal growth. As the groove at the re-entrant angle between the two parts provides a permanent step during crystal growth, twin crystals may grow into large crystals in this direction (Fig. 2.17c). In some cases, many twins are repeated at a microscopic scale, which can only be detected under a microscope.[4]

Figure 2.18a shows the "butterfly" twins of $BaTiO_3$. Here, the faces of plates correspond to the cubic {100} faces, whereas the twinning plane is parallel to {111}. Not to be confused with twin crystals are those of parallel growth, for which two examples are shown in Fig. 2.18b, c. We can see that all faces of the same crystal form are similarly oriented in parallel growth, whereas they are in reverse or mirror-image positions in twin crystals. If multiple crystals are joined with no special orientational relationship, they are neither twins nor those of parallel growth; they are simply called intergrown crystals.

Flux inclusion is another type of imperfection (Fig. 2.19a). It occurs when growth proceeds at a rate much faster than it is possible for flux particles to diffuse away from the surface. Accordingly, this condition is most likely to be met when supersaturation is too high. It occurs particularly when the spaces between the arms of dendrites are being filled, trapping the flux inside. Flux inclusion may be avoided by lowering the cooling rate or by using a less viscous flux.

[4]Crystals that have undergone structural transformation or strong stress can also show microscopic twins. These are called transformation and deformation twins, respectively, and are to be distinguished from the growth twins discussed here.

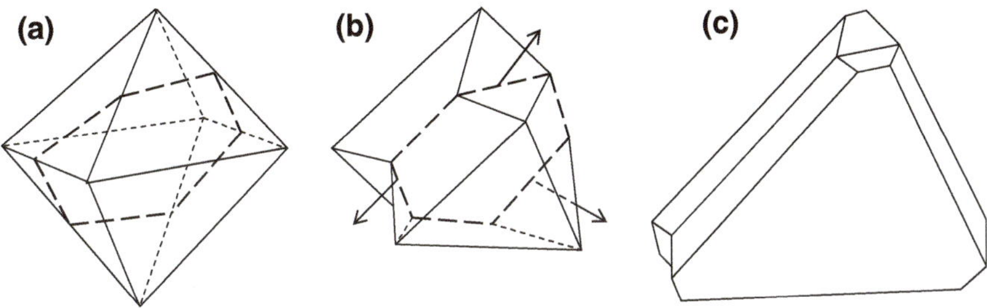

Fig. 2.17 **a** Octahedral crystal. **b** Twinned octahedral crystal, with arrows pointing from the re-entrant angles. **c** Tabular twinned crystal due to favored growth at re-entrant angles

Fig. 2.18 **a** Butterfly twins of $BaTiO_3$. **b** Parallel growth of cubic $PbMg_{1/3}Nb_{2/3}O_3$–$PbTiO_3$ crystals. **c** Parallel growth of octahedral $Y_2Ti_2O_7$ crystals

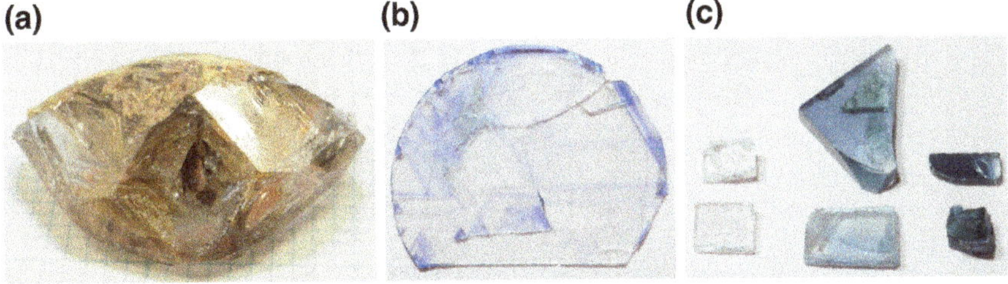

Fig. 2.19 **a** $Y_3Al_5O_{12}$ crystal with massive flux inclusions. **b** Blue-colored growth bands in $Al_{2-x}(Fe,Ti)_xO_3$ crystal. The size of this cracked crystal was limited by the crucible. **c** Color centers in $KTaO_3$ crystals. The blue color, which is caused by oxygen vacancies, increases in intensity from the crystals on the left to those on the right

When the supply of solute to the growing surface is interrupted several times during crystal growth, growth bands can form in the crystal. Growth bands are easily recognized when the crystal is transparent, and their shape often reflects the underlying symmetry of the crystal (Fig. 2.19b). As growth bands are spatial variations in composition or impurities, they can be prevented by minimizing the irregularities in growth conditions, such as temperature fluctuations. When many growth bands appear in a periodic sequence, they are usually referred to as growth striations. Often, growth striations along different crystallographic directions meet at the growth sector boundary; growth sectors are produced when different crystallographic directions grow with a different manner of defect formation.

There are several types of imperfection at the atomic level, which are called point defects. Vacancies, accidental insertion of atoms at interstitial positions, and substitution of impurity atoms are some of the examples. The number of point defects is generally decreased when crystal growth takes place at lower temperatures. When point defects produce color in otherwise colorless crystals, they are often called color centers (Fig. 2.19c).

In addition to screw dislocation, edge dislocation is another type of line defect that can provide active growth centers. An edge dislocation is formed where a plane of atoms extends only part of the way into a crystal lattice; it is shown as an inverted T in Fig. 2.20. In the figure, a series of edge dislocations of similar orientation makes the top half of the crystal slightly wider than the bottom half. This results in angular misorientation between the left half and the right half of the crystal, creating a low-angle grain boundary.

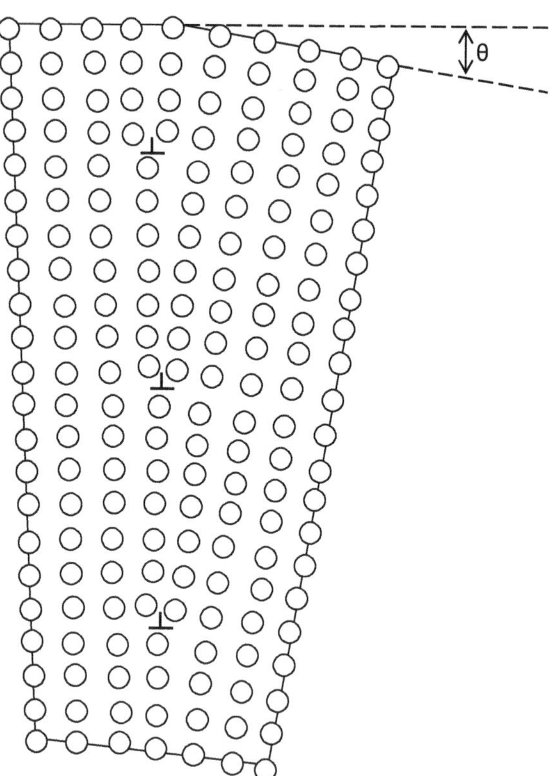

Fig. 2.20 A row of edge dislocations creating a low-angle grain boundary

We have looked at some of the most common types of imperfection found in flux-grown crystals. Some physical properties, such as the resistivity of semiconductors, are very sensitive to many types of imperfection, whereas others are more tolerant. Although imperfections are only viewed as a nuisance for the user of crystals, they can provide important insights on the growth mechanisms, allowing the crystal grower to improve growth conditions.

References

1. I. Sunagawa, *Crystals: Growth, Morphology and Perfection* (Cambridge University Press, Cambridge, 2005)
2. D. Elwell, H.J. Scheel, *Crystal Growth from High-Temperature Solutions* (Academic Press, London, 1975)
3. J.D.H. Donnay, D. Harker, Am. Min. **22**, 446–467 (1937)
4. P. Hartman, W.G. Perdok, Acta. Cryst. **8**, 49–52, 521–524, 525–529 (1955)
5. K. Watanabe, S. Tsunoda, J. Cryst. Growth **79**, 953–962 (1986)
6. S.T. Jung, D.Y. Choi, S.J. Chung, J. Cryst. Growth **160**, 305–309 (1996)
7. F.C. Frank, Disc. Faraday Soc. **5**, 48–54 (1949)
8. W.K. Burton, N. Cabrera, F.C. Frank, Proc. R. Soc. Lond. A **243**, 299–358 (1951)
9. C.T. Lin, Physica C **337**, 312–316 (2000)
10. I. Nakada, R. Akaba, T. Yanase, Jpn. J. Appl. Phys. **11**, 1583 (1972)

Chapter 3
Phase Diagrams for Flux Growth

The beginner is best advised to start flux growth by following known recipes or procedures. It is important for a person who has not performed any crystal growth to get the "feel" of it, and this is most easily accomplished by reproducing somebody else's success. However, there soon comes a time when it is necessary to perform original experiments—because known recipes will not produce crystals of the required size and/or quality, or because there is a need to obtain new crystals. This is when knowledge of phase diagrams and fluxes becomes important. From a relevant phase diagram, it is possible to obtain crucial information such as the appropriate starting composition, temperature profile, and cooling rate. Phase diagrams also give a better understanding of the growth processes. This chapter describes the use of phase diagrams for flux growth, whereas the next chapter will focus on the choice of flux.

3.1 Introduction

When single crystals of a compound grow from solution, the compound undergoes a phase transformation: it changes from being a part of a liquid to being a solid. This transformation is usually brought about by a change in temperature, but it can also happen following a change in composition or pressure. Phase transformations under various conditions of temperature, composition, and pressure are graphically represented in a phase diagram. To put it another way, a phase diagram shows which phase or mixture of phases is thermodynamically stable (that is, in equilibrium) under certain conditions of temperature, composition, and pressure. A useful phase diagram therefore reveals the proper methods and procedures for growing the desired crystals. It also explains why some growth experiments would not work. For many flux growth experiments, the gaseous phases and effects of

M. Tachibana, *Beginner's Guide to Flux Crystal Growth*, NIMS Monographs,
DOI 10.1007/978-4-431-56587-1_3

pressure can be ignored for practical purposes. This means that we consider the liquid–solid phase diagrams at ambient pressure (1 atm) of, typically, air for oxide systems and an inert atmosphere for non-oxide systems.

3.2 Simple Eutectic System

To understand the use of phase diagrams, let us start with the general case shown in Fig. 3.1. Here, temperature is plotted on the vertical axis, where the bottom is still much higher than room temperature. The horizontal axis gives the compositions of solute and flux, with the scale from left to right referring to increasing percentage of flux in the system. The composition may be given in either mole percent or weight percent. The flux can be an element, a compound, or a combination of compounds. Because this type of phase diagram is not limited to uses in flux growth, the solute is also labeled as component A and the flux as component B. The diagram shows that they melt congruently (that is, without decomposition) at T_A and T_B, respectively. The presence of two components in the system makes it a binary phase diagram; however, the term "pseudo-binary" is more appropriate if one or both of the components can be broken down into further components (such as when a combination of compounds is used as the flux).

In the region marked "liquid" at high temperatures, any combination of A and B becomes a homogeneous liquid—the system has the same chemical and physical properties everywhere, and it is not possible to mechanically separate A from B. On the other hand, the low-temperature region marked "A + B" is a heterogeneous solid system. Here, solid A and solid B coexist as separate entities, such that physical properties and composition vary in different parts of the system. At intermediate temperatures, there are two separate regions of partial melting, which are also heterogeneous in character. The left region, marked "A + liquid", is a mixture of solid A and liquid of different composition. In the right region marked "B + liquid", solid B coexists with liquid of different composition. The curves

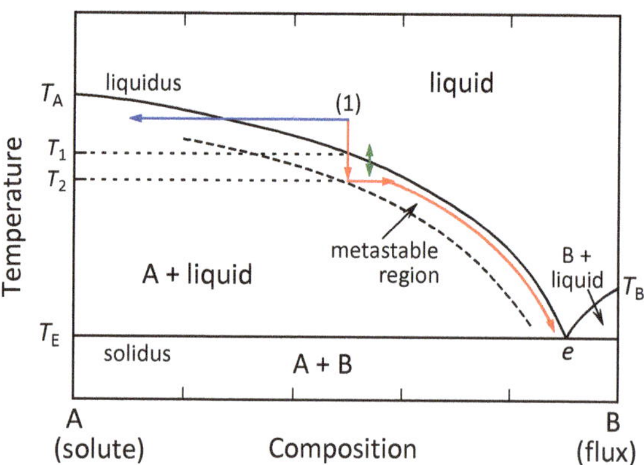

Fig. 3.1 Binary phase diagram of a simple eutectic system. The metastable region shown here is schematic; it may not have a constant width in real cases

separating "liquid" from "A + liquid" and "B + liquid" are called liquidus curves. The horizontal line separating "A + liquid" and "B + liquid" from "A + B" is called the solidus. The liquidus curves and the solidus line meet at a single point marked "*e*", which is the eutectic point. It is only at the eutectic point that the liquid, solid A, and solid B coexist in equilibrium. This phase diagram is an example of a simple eutectic system.

If we look at the left liquidus curve that stretches from the melting point of pure A to the eutectic point, we can see how the addition of B (flux) lowers the melting temperature of A (solute) from T_A to T_E. This shows how a flux can lower the crystallization temperature of the solute. The liquidus curve also defines the temperatures and compositions at which A as a solid and as a constituent in liquid are in equilibrium. In other words, the liquidus curve separates the region of homogeneous liquid from the region of solid solute plus liquid, just as the solubility curve (*see* Sect. 2.2) separates the region of unsaturated solution from that of supersaturated solution. A liquidus curve and a solubility curve are really the same thing plotted in a different context, and this is why they have a similar shape.

To understand the process of flux growth, let us follow a particular path in the phase diagram. We begin with point (1), which is a homogeneous liquid of 50% solute and 50% flux. (To begin at this point, it is assumed that the starting solid mixture has been heated and well equilibrated at this temperature.) If this liquid is cooled slowly, it moves vertically downward in the "liquid" region of the phase diagram. Once the temperature falls below the liquidus curve, the solid solute should start to appear in the liquid. However, because some additional energy is needed for nucleation, crystallization does not begin until T_2; for this reason, the region between the liquidus curve and the dashed curve is called a metastable region. The width of the metastable region corresponds to how much the solution undercools or supersaturates before nucleation occurs spontaneously.

Most phase diagrams do not show metastable regions, simply because metastable states are not thermodynamically stable and do not really belong in equilibrium phase diagrams. Metastable regions are also difficult to determine experimentally and different experiments may give different results. Nevertheless, for the purpose of flux growth, the presence of a reasonably large metastable region is important for reducing multi-nucleation and producing large crystals.

When crystal growth finally starts at T_2, it reduces the degree of supersaturation by removing some of the excess solute dissolved in the liquid. The system therefore approaches equilibrium, as shown by the horizontal arrow at T_2. This process is more generally explained as follows: In the region where two phases exist ("A + liquid" in this case), the compositions of two phases in equilibrium are given by the intersection of an isothermal line with the phase boundaries. Using this rule, the composition of the liquid phase is obtained as the position on the liquidus curve indicated by the horizontal arrow, while the composition of the solid is simply A.

As the solution continues to cool slowly below T_2, the presence of crystals already formed in the solution allows growth to continue at much lower supersaturation. This means that the system does not deviate much from equilibrium, such that the temperature dependence of the liquid composition can be plotted as the curved arrow down to the eutectic point. At the temperature immediately above the eutectic point, the liquid has a composition of 10% solute and 90% flux. Below the eutectic point, this remaining liquid solidifies around the crystals that have already been grown.

In the above example, the composition of the liquid starts with 50% solute and ends with 10% solute at the eutectic point. Since the composition of the entire system must remain constant (we presume no evaporation loss in this case), the 40% loss of solute in the liquid corresponds to the amount of solute that becomes crystals. Accordingly, more solute can be crystallized if the growth starts with, say, 55% solute and 45% flux. However, in order to start with a richer solute composition, we have to go to higher temperatures to dissolve the solute completely. There are several possible reasons why this is inconvenient and so we must start with the solute at 50% or lower: (1) the solute or flux becomes increasingly volatile at higher temperatures, (2) the upper temperature is limited by the furnace or crucible used in the growth, (3) the liquidus curve has an unfavorable slope at higher temperatures (see below), and (4) the solute transforms into another structure at higher temperatures, although this possibility is not explicitly shown in Fig. 3.1.

From Fig. 3.1, we can see that flux growth occurs when the solution moves down the liquidus curve of the phase diagram. A large liquidus curve is therefore a desirable feature in flux growth. In addition, the liquidus curve should have a suitable slope. If the slope is too shallow, a small change in temperature would cause a large change in liquid composition. This can lead to instability during growth, resulting in crystals with many defects and compositional variations. On the other hand, a steep liquidus curve would not produce large crystals, because only a small change in composition is available for crystal growth.

Single crystals can also be grown by the evaporation method or temperature gradient method. In the evaporation method, a volatile flux is evaporated off the solution to induce supersaturation. This process is shown in Fig. 3.1 as the horizontal arrow pointing left from (1). Because crystals can grow at constant temperature, the evaporation method is useful when the available range of growth temperatures is limited. However, crystallization often occurs at the surface of the solution, because of the high local supersaturation created by the flux evaporation. It is also difficult to control growth rate, and the resulting crystals tend to contain many defects. Another problem is possible damage to the furnace from the evaporated flux.

In the temperature gradient method, the solution is placed in a temperature gradient shown by the vertical arrow pointing up and down in Fig. 3.1. During the experiment, the undissolved solute is placed at a temperature above the liquidus curve, while crystal growth occurs at a temperature below the liquidus. As the undissolved solute is dissolved slowly into the solution, it is carried to the cooler region and incorporated into the growing crystal. This method usually requires a seed crystal to be placed in the growth region, and it demands much effort in finding the appropriate growth conditions. However, this technique can produce large and high-quality crystals.

3.3 Examples of Eutectic Systems

We can now examine specific examples for flux growth. Figure 3.2a shows the phase diagram of $BaTiO_3$–KF [1]. The cubic $BaTiO_3$ is an important compound that becomes ferroelectric at room temperature. $BaTiO_3$ has a congruent melting point of around 1600 °C, but cooling the melt results in the hexagonal phase of this compound. It is not possible to obtain good crystals of cubic $BaTiO_3$ by cooling the hexagonal crystals, because the large difference in crystal structure destroys the crystals at the phase transformation. It is therefore necessary to grow cubic $BaTiO_3$ directly from liquid. This can be done by using KF as a flux, which lowers the growth temperature to the region where the cubic phase becomes the primary crystallizing phase. In practice, the high volatility of KF limits growth temperatures to below 1200 °C. This method produces the butterfly twins of $BaTiO_3$ crystals, as discussed in Sect. 2.3. (Commercially, large and high-quality crystals of $BaTiO_3$ are grown by a seeded technique using TiO_2 as a self-flux.)

The phase diagram of the system $BaFe_2As_2$–FeAs is shown in Fig. 3.2b [2]. The crystal growth of $BaFe_2As_2$ must be performed in an oxygen-free environment, such as in a sealed silica glass tube. Because the melting point of $BaFe_2As_2$ is above the working temperature of silica glass, FeAs is used as a self-flux to lower the growth temperature. High-quality single crystals of $BaFe_2As_2$ and related compounds can be grown by this technique.

Figure 3.2c shows the phase diagram of PbO–TiO_2 [3] for the crystal growth of $PbTiO_3$. Unlike the examples discussed so far, this phase diagram shows the desired compound as a combination of two components. However, it can be seen that the phase diagram is simply a combination of two eutectic systems, one in the composition range of PbO–$PbTiO_3$ and the other in the range of $PbTiO_3$–TiO_2. Although $PbTiO_3$ can be grown from its own melt at 1285 °C, the high vapor pressure of PbO at this temperature hinders the growth of high-quality crystals. The phase diagram suggests that $PbTiO_3$ can be grown by using either excess PbO or TiO_2 as a flux, and that PbO should be a better choice because of its wider liquidus curve. Incidentally, this example highlights an important feature of flux growth: whereas the melt technique requires the precise stoichiometric control of the liquid composition, the flux method is much more forgiving in the sense that growth can take place over a wide variation of compositions, and evaporation loss is tolerated. In practice, PbO alone [4] or PbO and B_2O_3 [5] can be used as the flux to grow $PbTiO_3$ crystals.

Finally, Fig. 3.2d shows the phase diagram of PbO–PbF_2 [6]. Both PbO and PbF_2 are each used as a flux for many oxide systems. This phase diagram shows that the melting temperature is lowered when PbO and PbF_2 are combined, down to the eutectic point of 490 °C. Using a low-melting flux is often advantageous, as it can reduce growth temperatures and increase solubility. For such reasons, in flux growth a combination of two or more compounds is often used as the flux. Chapter 4 will describe many such examples.

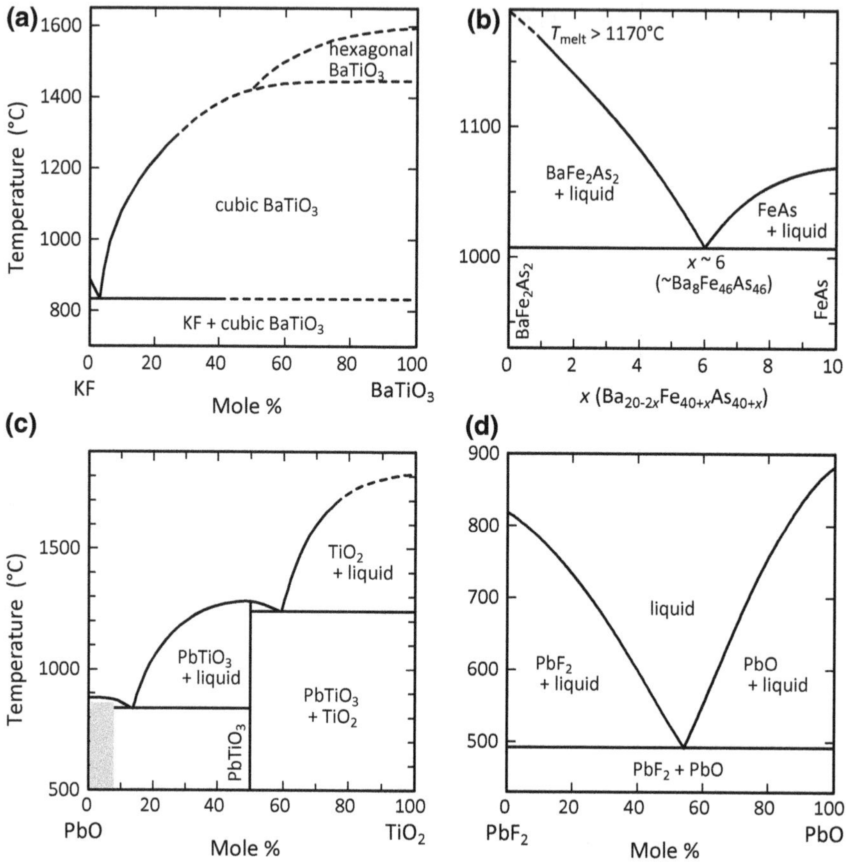

Fig. 3.2 **a** $BaTiO_3$–KF phase diagram (from [1]). **b** $BaFe_2As_2$–FeAs phase diagram (from [2]). **c** PbO–TiO_2 phase diagram (from [3]; complex behavior near PbO is omitted). **d** PbO–PbF_2 phase diagram (from [6])

3.4 Incongruently Melting Compounds

Some compounds do not melt into a liquid of the same composition; instead, they decompose into another solid and a liquid. This is called incongruent melting, and Fig. 3.3 gives such an example. Here, the compound A_2B melts incongruently at T_1, turning into a mixture of solid A and liquid of composition p. To understand why A_2B shows incongruent melting, it is useful to think of what would happen if A had a lower melting point at T_{A*}. This is shown as dashed curves in the figure, and we can see that A_2B now melts congruently. Therefore, incongruent melting can be considered to be a result of one component having a high melting temperature.

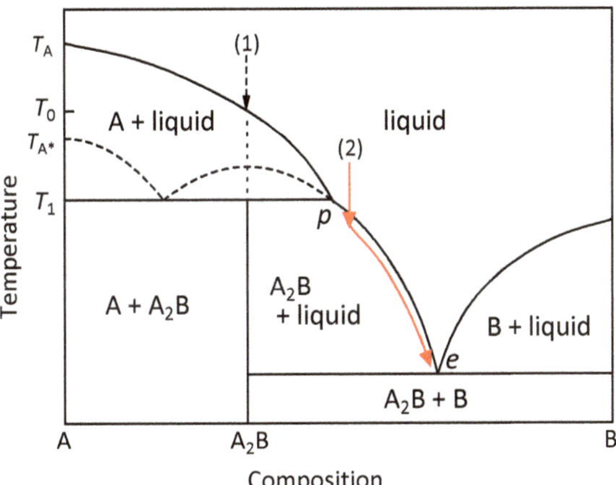

Fig. 3.3 Binary system with incongruently melting compound A_2B

If a compound melts incongruently, it becomes difficult to grow single crystals by cooling its own melt. This is explained as follows: When a liquid of composition A_2B is cooled, as shown by the dashed arrow starting at (1), crystals of A would start to appear below T_0. By the time it reaches T_1, a significant amount of A has been crystallized and the liquid now has composition p. Since A_2B is the thermodynamically stable phase below T_1, solid A and the liquid (with composition p) should react to form A_2B. However, as the reaction can occur only at the surface of the crystals of A, the reaction is slow and usually stops before reaching completion. What is then produced on further cooling is a solid mixture of A_2B, A, and B, rather than single crystals of A_2B. The point "p" is called the peritectic point, and this phase diagram is an example of a peritectic system. The peritectic point is the only place where A, A_2B, and liquid can coexist in thermodynamic equilibrium.

Single crystals of A_2B can be grown by adjusting the liquid composition such that A_2B becomes the primary solid phase to appear on cooling. This occurs between the compositions of point "p" and point "e" in the phase diagram; a possible path of flux growth is shown by the solid arrows starting at (2). This is basically the technique of using B as a self-flux to grow single crystals of A_2B. This phase diagram also helps to understand why the low-melting-point component (B in this case) is often the one used as a self-flux.

If the liquidus curve between point "p" and point "e" is limited in composition and/or temperature such that the use of B as a self-flux is not feasible, a different flux can be used to grow single crystals of A_2B. If the new flux is labeled as C, we are now dealing with a three-component or ternary system of A–B–C. An equilateral triangle can be used to show the composition, with each corner representing 100% of one component. For the temperature scale, another axis can be added to the triangle. However, it is difficult to plot and visualize such a three-dimensional diagram, so only a triangle is used in many cases to discuss the phases present at a particular temperature (if necessary, temperature variations can be plotted with contour lines in the triangle). Figure 3.4a, b show two possible outcomes of adding a new flux C [7]. Here, the temperature is above the melting point of C, and A, A_2B, and B are the same as in Fig. 3.3.

In Fig. 3.4a, b, the straight lines that extend from A_2B to C represent all the possible combinations of A_2B and C, with A and B fixed at the 2:1 ratio. In the case of Fig. 3.4a, the solubility limit (the dashed line from 2 to 3) intersects with the A_2B stoichiometry line. This is called congruent saturation, where crystals of A_2B can directly form from a solution that is saturated with respect to stoichiometric A_2B. A different scenario is found for incongruent saturation, which is shown in Fig. 3.4b. In this case, attempts to crystallize A_2B from a solution containing the stoichiometric 2:1 ratio would first lead to the formation of solid A. This process is continued until the liquid composition is shifted to point 2, where solid A_2B now starts to appear. To avoid the formation of solid A, the initial solution must be saturated with A and B in some ratio that is richer in B than A_2B. Another way to avoid the formation of solid A is to find a different flux, one that will lead to congruent saturation. Usually, the best strategy of finding a congruent saturating flux is to look for a flux that has a high solvent power at as low a temperature as possible. Note that incongruent saturation is not directly related to incongruent melting, and congruently melting compounds can show incongruent saturation under certain fluxes.

An example of an incongruently melting compound is La_2CuO_4, which is found in the phase diagram of La_2O_3–CuO [8] (Fig. 3.5a). As expected from the large liquidus curve, single crystals of La_2CuO_4 can be grown by using CuO as a self-flux. The phase diagram also shows that the growth must be stopped above 1050 °C, before $La_2Cu_2O_5$ becomes the stable phase. Although other fluxes such as PbO [9] and $Li_4B_2O_5$ [10] have been used to grow La_2CuO_4, the best crystals are obtained when CuO is used as the self-flux.

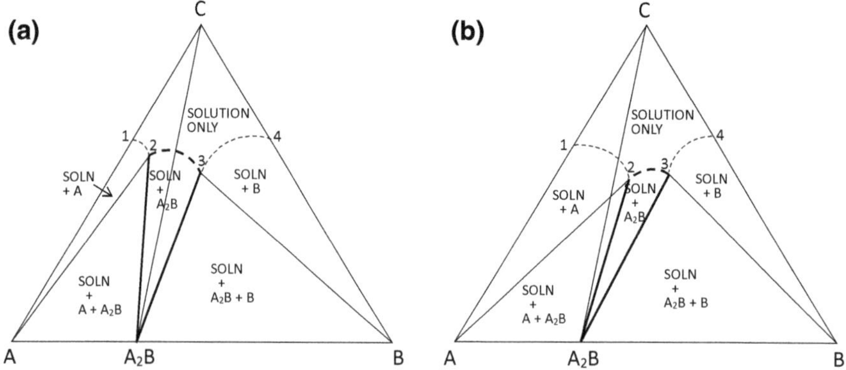

Fig. 3.4 **a** Ternary system A–B–C with the compound A_2B congruently saturating in C. **b** Ternary system A–B–C with A_2B not congruently saturating in C

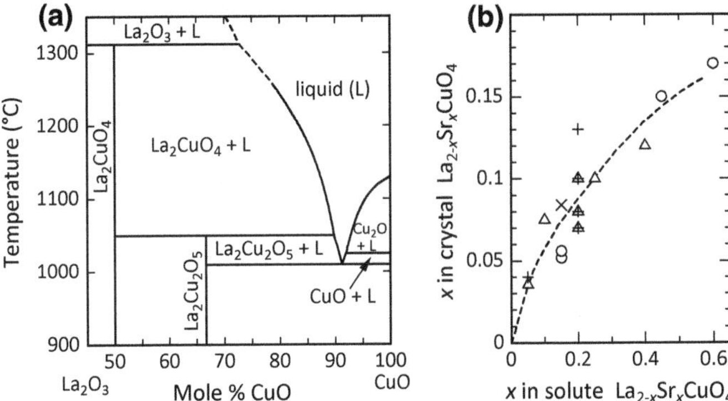

Fig. 3.5 **a** La_2O_3–CuO_2 phase diagram (after [8]). **b** Dependence of Sr content in $La_{2-x}Sr_xCuO_4$ crystals on Sr content in the starting material (after [11]; different symbols indicate results from different studies)

La_2CuO_4 becomes a superconductor when Sr^{2+} ions replace some of the La^{3+} ions. The resulting composition, $La_{2-x}Sr_xCuO_4$, is an example of a solid solution, for which the maximum superconducting transition temperature of 40 K is found at $x = 0.15$. By replacing some La_2O_3 in the starting material with SrO ($SrCO_3$ is used in practice; it loses CO_2 on heating), single crystals of $La_{2-x}Sr_xCuO_4$ can be grown in a manner similar to La_2CuO_4. However, one problem is that Sr^{2+} ions in the solution are less willing than La^{3+} ions to go into the crystal, partly because the electric charge is different. This situation can be judged from the data [11] given in Fig. 3.5b, which shows that in order to grow crystals of $La_{2-x}Sr_xCuO_4$ with $x = 0.15$, the composition in the solution must be close to $x = 0.45$. To understand this and other complexities of solid solutions, we must next look at their phase diagrams.

3.5 Solid Solutions

Thus far, we have looked at phase diagrams in which solids always have a fixed composition. Solid solutions, in turn, have a solid phase whose composition can change continuously. The simplest case of solid solutions is shown in Fig. 3.6; here, it occurs in the entire range of $A_{1-x}B_x$, from $x = 0$ to $x = 1$. Such a phase diagram is found for $KTa_{1-x}Nb_xO_3$, which forms a complete series of solid solutions between $KTaO_3$ and $KNbO_3$ [12]. In Fig. 3.6, sandwiched between the high-temperature liquid phase and the low-temperature solid-solution phase is the two-phase region, where a mixture of solid solution and liquid exists. Both the liquidus and solidus move downward as B is added to A, or looked at from the other way, they move upward as A is added to B.[1]

On cooling a liquid of composition y from point (1), crystallization starts at T_1 if supercooling is ignored. Just as we saw in the case of a eutectic system, the compositions of the crystallized solid and the coexisting liquid can be determined by drawing an isothermal line: at T_1, the crossing of this line with the solidus curve gives the composition of the solid solution as x, while the crossing of this line with the liquidus curve gives the liquid composition as y. At this point, the fraction of the solid solution in the system is still close to zero. When the temperature is further lowered to T_2, the fraction of the solid solution in the system is increased, but its composition is now changed to x', while the composition of the liquid is now at y' (the fraction of the solid solution in the system is given by $yy'/x'y'$). The fractional and compositional changes in the solid solution and liquid will continue down to T_3, at which the system is nearly completely crystallized; the crystal now has the composition y, while the last remaining liquid has the composition y''.

This example shows that the crystallizing composition changes continuously when a solid solution is cooled from its melt. Although a uniform crystal of solid solution is more thermodynamically stable than a non-uniform crystal, it is usually very difficult for the atoms in the crystal to diffuse into their equilibrium positions. As a result, the crystal often contains a compositional gradient, with the inner region being richer in A and the outer zone richer in B.

In flux growth, single crystals of uniform solid solutions can be obtained by using a suitable flux. To illustrate this point, Fig. 3.7 shows a case where A and B form a continuous series of solid solutions, with the flux C forming simple eutectics with both A and B [7]. As we just saw above, if a solid solution of composition x is to be grown from a melt, the liquid must have composition y. On the other hand, if C is used as the flux, the phase diagram shows that single crystals of a uniform composition x can be grown by simply cooling the solution along the curve from T_1 to T_2. Even when such an ideal situation does not occur, the use of a flux often helps to reduce the degree of inhomogeneity.

The formation of solid solutions over the entire composition is observed only when the components are very similar in size and chemical bonding properties. It is more common for solid phases to be only partially soluble to each other, as in the case shown in Fig. 3.8a; here, partial solubility is found in a eutectic system. In the region marked "A_{ss}", component B is dissolved in crystals of A to form a solid solution of $A_{1-x}B_x$. The extent of solid solubility varies with temperature, and the maximum occurs at the eutectic temperature. Similarly, in the region marked "B_{ss}", component A dissolves into B to form a solid solution of $B_{1-x}A_x$. If B is used as a flux to grow single crystals of A, the phase diagram predicts that the flux will be

[1]In some systems, the solid solution between the two end members becomes unstable, or immiscible, at low temperatures. This is represented in the phase diagram as a domed region that extends down to 0 K, and results in the separation of the solid into two solid phases of different composition, usually at a microscopic scale.

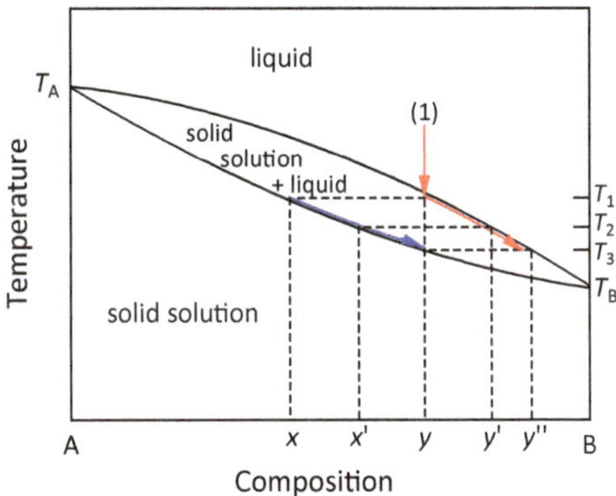

Fig. 3.6 Simple binary system forming solid solutions in the entire composition range

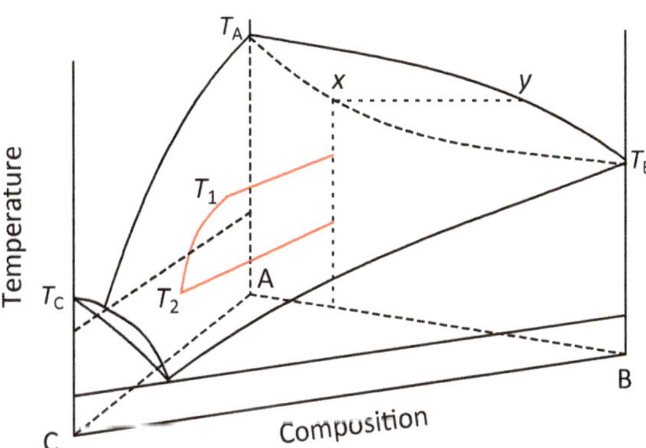

Fig. 3.7 Simple ternary diagram with two components A and B forming a complete solid solution range and each forming a simple eutectic system with the third component C

incorporated into the lattice of the crystals. Obviously, B is not a good flux for growing pure crystals of A. In many cases, it is the incorporation of flux into the crystals that prevents the application of a particular flux.

In phase diagrams, compounds are often plotted as vertical lines (*see* Fig. 3.8b for a congruently melting compound AB). This implies that they have a strictly fixed, stoichiometric composition, where the constituent atoms are present in a simple ratio, such as 1:1 for AB and 1:1:3 for ABC_3. This is often an oversimplification, because many compounds can exist as a single phase over a certain range of composition; *see* Fig. 3.8c for compound AB. The extent of such off-stoichiometry varies widely among different types of compounds, usually being small in ionic compounds with noble-gas electron configurations. Because even a small variation in composition can affect physical properties, it must be treated in a manner similar to impurities.

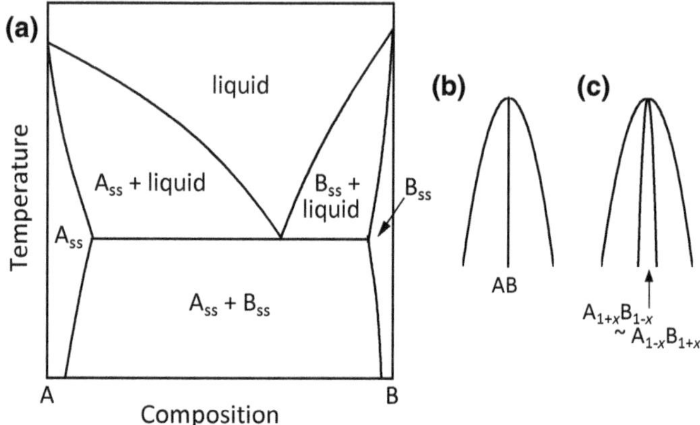

Fig. 3.8 **a** Eutectic binary system showing partial solid solubility of the end members. **b** Stoichiometric line compound AB. **c** Area around the compound AB showing an extended solid solubility region

3.6 Oxygen Partial Pressure and Oxidation State

When oxide crystals are grown in air, the metal ions usually have the "normal" oxidation (valence) states—meaning stable oxidation states under an oxygen partial pressure of 0.21 atm, that is, the value in air. If crystals are grown under different oxygen partial pressures, different oxidation states may become more stable. Such modification is especially important for the first-row transition metals, for which the following states are widely found: Ti^{2+}, Ti^{3+}, Ti^{4+}; V^{2+}, V^{3+}, V^{4+}, V^{5+}; Cr^{3+}, Cr^{4+}; Mn^{2+}, Mn^{3+}, Mn^{4+}; Fe^{2+}, Fe^{3+}, Fe^{4+}; Co^{2+}, Co^{3+}; Ni^{2+}, Ni^{3+}; Cu^{+}, Cu^{2+}. The presence of multiple oxidation states also allows oxygen non-stoichiometry to be easily formed in some cases, as the variation in oxygen content can be charge balanced by changing the average oxidation state. To understand the relationship between oxidation states and oxygen partial pressure, we must look at phase diagrams in which oxygen partial pressure is a variable.

Figure 3.9a is a phase diagram for the Cu–O system [13], which shows the stable solids as a function of temperature and oxygen partial pressure. At ambient temperature and pressure, CuO (with Cu^{2+}) is the thermodynamically stable state. Above 700 °C, CuO becomes relatively unstable in air and gradually loses O_2 to form Cu_2O (with Cu^{+}). Above 1200 °C, Cu_2O tends to dissociate into metallic copper (Cu^0) and O_2. Under reduced oxygen partial pressures, these two steps of decomposition occur at lower temperatures; alternatively, increasing the oxygen pressure stabilizes the high oxidation state. Similar behavior is observed for the Mn–O system [14] shown in Fig. 3.9b and for other metal–oxygen systems in general. These properties of simple oxides provide rough ideas for the relative stabilities of different oxidation states in more complex oxides.

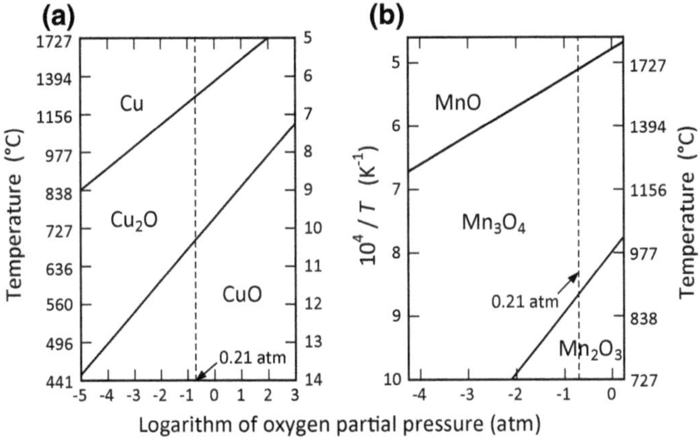

Fig. 3.9 **a** Cu–O phase diagram showing the stable solids as a function of temperature and oxygen partial pressure (after [13]). **b** Mn–O phase diagram (after [14])

On the basis of thermodynamics, oxides with unusually high oxidation states can be grown under very high oxygen pressures. However, such experiments require expensive equipment to contain the necessary high pressures. One way around this problem is to adjust the acid–base chemistry of the flux solution, which can influence the oxidation state without changing the pressure [15]. For example, the oxidizing flux NaOH can stabilize oxides with unusually high oxidation states, such as $Ba_3Mn_2O_8$ (Mn^{5+}) and Ba_2NaOsO_6 (Os^{7+}) [16].

On the other hand, low oxygen partial pressures can be realized more easily. To obtain such a condition for stabilizing low oxidation states, a flow of controlled gases is introduced into the growth environment. If the required oxygen partial pressure is above 10^{-5} atm, a mixture of Ar/O_2 or N_2/O_2 is often used. For lower oxygen partial pressures, buffer gases such as CO/CO_2 or H_2/H_2O can be used. (The controlled reaction of $CO_2 = CO + \frac{1}{2}O_2$ or $H_2O = H_2 + \frac{1}{2}O_2$ produces the required oxygen partial pressures.) In some cases, it is more convenient to fix the total oxygen content of the system. For example, if the growth is carried out in a vacuum-sealed silica glass tube, there is no oxygen coming into or going out of the system. In this case, the oxidation state of the starting material can be maintained throughout the growth, provided that neither the flux nor the crucible influences the oxidation state. Single crystals of $LiVO_2$ (V^{3+}) [17] and LiV_2O_4 (average: $V^{3.5+}$) [18, 19] have been grown in this way, as higher oxidation states are more stable in air.

3.7 Determination of Phase Diagrams

We have looked at some of the major types of phase diagrams. The thermodynamic foundations of phase diagrams, which we have not discussed in any length, can be found in textbooks on physical chemistry, ceramics, or metallurgy. Many new phase diagrams are reported each year and there are a number of books that catalog the published phase diagrams. The well-known series for oxides and salts is *Phase Equilibria Diagrams*, formerly known as *Phase Diagrams for Ceramists* [20]. The phase diagrams for many of the mixtures of compounds used as the flux, which will be described in Chap. 4, can be found in this series. As for phase diagrams on alloys and intermetallic compounds, there are books by Hansen [21], Elliot [22], Shunk [23], Moffatt [24], and Massalski [25], as well as more recent ones such as the one by Okamoto [26]. Although these references contain thousands of phase diagrams, it is quite likely that the one needed for the next flux growth is missing. In such a case, some form of preliminary experiments is often performed by the crystal grower.

Perhaps the quickest way to obtain the necessary information is to conduct trial growth experiments—choose a convenient maximum temperature, and cool a number of crucibles containing mixtures of various solute:flux ratios. The first run usually covers a large composition region, and once a promising region is found, the second run can be focused on that region by using smaller composition intervals. By using fast cooling rates and small crucibles, the growth condition can be optimized fairly quickly and at a low cost.

3.7.1 Quenching Method

In some cases, it is desirable to learn more about the phase diagram than can be found from trial growth experiments. In particular, any information on the primary crystallizing field and liquidus curve is valuable (even when a mixture of compounds is used as a flux, the system should behave as pseudo-binary in a certain region). A basic method of making a phase diagram is to quench the solutions from various temperatures. In this technique, a uniform mixture of solute and flux is sealed in a container and then heated at the target temperature until equilibrium is established. The container is then rapidly quenched by dropping into water or liquid nitrogen. When quenched, any solid phase that was present at the high temperature is retained, whereas phases that were liquids become glasses. These phases can be identified by microscopy and X-ray diffraction analysis.

If the quenched sample contains more than one crystalline phase, a quenching experiment of the same composition is repeated at a higher temperature. If this sample now contains only one crystalline phase, this phase is the primary crystallizing phase for that part of the system. The liquidus temperature can be determined by repeating the experiment at still higher temperatures, until the primary phase disappears and only a glass is found in the quenched sample. Figure 3.10a shows an example of a phase diagram determined by this method [27].

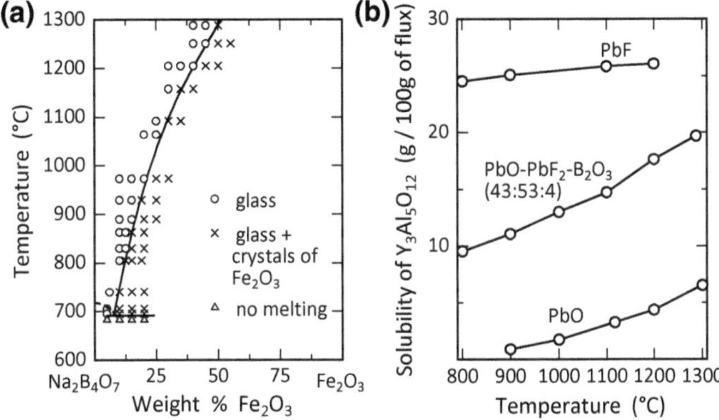

Fig. 3.10 **a** Fe_2O_3–$Na_2B_4O_7$ phase diagram obtained from quenching experiments (after [27]). **b** Solubility of $Y_3Al_5O_{12}$ in three different fluxes (after [28]). Note that the solubility is large in PbF_2, but the small temperature dependence makes it unsuitable as a flux for slow-cooling growth

3.7.2 Solubility Determination

As the quenching method requires several experiments to determine a single point of the liquidus curve, it can be time consuming. If the desired compound is available in the form of small crystals or hard polycrystalline pieces, the liquidus curve (or solubility curve) can be determined by measuring the solubility at different temperatures. This is carried out by soaking the sample in a given amount of flux at fixed temperatures, and then quenching it. By dissolving away the flux and measuring the weight loss of the sample, its solubility at different temperatures can be determined. To avoid any loss from evaporation, it is often necessary to seal the container. For experiments on oxide systems, platinum tubes can be sealed by squeezing them flat at the ends and welding with an arc welder or oxygen–propane torch. Figure 3.10b shows the solubility curves of $Y_3Al_5O_{12}$ in three different fluxes [28].

3.7.3 Hot-Stage Microscopy

With the use of a heating stage (often called a hot stage), the process of melting and crystallization can be observed under a microscope. On warming a uniform solid mixture of solute and flux, the first hint of melting at the solidus temperature comes from the gentle rounding of the edges. The amount of liquid increases on further heating, and the sample turns into a complete liquid at the liquidus temperature. To better distinguish the solid from liquid, polarized light is often used to enhance the contrast and to improve image quality. Although hot-stage microscopy becomes less reliable for samples that are volatile or optically opaque, it is a fast technique for determining the phase diagram of novel systems.

3.7.4 Differential Thermal Analysis

In differential thermal analysis (DTA), phase boundaries are determined by detecting the changes in heat content during phase transformation. In the experiment, a sample cell and an inert reference cell are placed next to each other in a furnace and heated or cooled at a constant rate (usually 10 °C per minute). Thermocouples measure the absolute temperature, as well as the temperature difference between the sample and reference material. When some thermal event such as melting or decomposition occurs in the sample, the sample temperature starts to differ from the reference temperature. This is recorded as an anomaly in the DTA curve as a function of temperature.

Figure 3.11 describes the results of DTA for a eutectic system. On heating composition (1), which is pure A, a peak appears at its melting point. Usually, the transition temperature is taken as the temperature at which deviation begins from the baseline, rather than the peak temperature. On heating compositions (2) and (3), a peak appears at the eutectic temperature, where the melting starts. This peak is superposed on a much broader anomaly, which arises from continuous melting and terminates at the liquidus temperature. The heating curve of composition (4) gives only a peak at the eutectic point, and the heating curve of composition (5) gives data on the part of the phase diagram where component B becomes the primary crystallizing phase.

As suggested from Fig. 3.11, it is sometimes difficult to determine the liquidus temperature by DTA. However, the technique can be used for volatile samples by

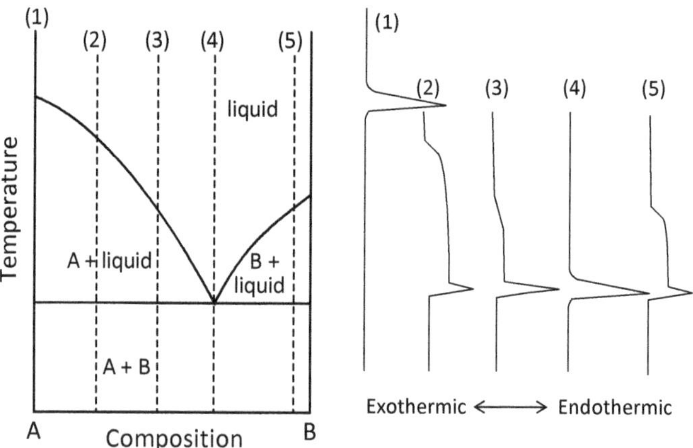

Fig. 3.11 Traces of differential thermal analysis (DTA) for a eutectic system

sealing the container cell, and the results on cooling curves provide information on the sample's supercooling behavior. Many differential thermal analyzers are capable of performing simultaneous thermogravimetry (measurement of changes in sample mass as a function of temperature), and are called TG-DTA.

References

1. C. Karan, B.J. Skinner, J. Chem. Phys. **21**, 2225 (1953); C. Karan, J. Chem. Phys. **22**, 957 (1954)
2. R. Morinaga, K. Matan, H.S. Suzuki, T.J. Sato, Jpn. J. Appl. Phys. **48**, 013004 (2009)
3. B. Jaffe, W.R. Cook, H. Jaffe, *Piezoelectric Ceramics* (Academic Press, New York, 1971)
4. B.N. Sun, Y. Huang, D.A. Payne, J. Cryst. Growth **128**, 867–870 (1993)
5. A. Kania, A. Slodczyk, Z. Ujma, J. Cryst. Growth **289**, 134–139 (2006)
6. E.M. Levin, C.R. Robbins, H.F. McMurdie, *Phase Diagrams for Ceramists* (The American Ceramic Society, Columbus, OH, 1964), p. 470
7. J.W. Nielsen, R. R. Monchamp, in *Phase Diagrams: Materials Science and Technology*, vol. 3, ed. by A.M. Alper (Academic Press, New York, 1970), pp. 2–52
8. A.N. Maljuk, A.B. Kulakov, G.A. Emel'chenko, J. Cryst. Growth **151**, 102–106 (1995)
9. S.-W. Cheong, Z. Fisk, R.S. Kwok, J.P. Remeika, J.D. Thompson, G. Gruner, Phys. Rev. B **37**, 5916–5919 (1988)
10. R.J. Birgeneau, C.Y. Chen, D.R. Gabbe, H.P. Jenssen, M.A. Kastner, C.J. Peters, P.J. Picone, T. Thio, T.R. Thurston, H.L. Tuller, Phys. Rev. Lett. **59**, 1329–1332 (1987)
11. C. Changkang, Prog. Cryst. Growth Charact. Mater. **24**, 213–267 (1992)
12. D. Rytz, H.J. Scheel, J. Cryst. Growth **59**, 468–484 (1982)
13. A.L. Pranatis, J. Am. Ceram. Soc. **51**, 182 (1968)
14. E.M. Levin, C.R. Robbins, H.F. McMurdie, *Phase Diagrams for Ceramists* (The American Ceramic Society, Columbus, Ohio, 1964), pp. 39–40
15. S.J. Mugavero III, W.R. Gemmill, I.P. Roof, and H-C. zur Loye, J. Solid State Chem. **182**, 1950–1963 (2009)
16. I.R. Fisher, M.C. Shapiro, J.G. Analytis, Philos. Mag. **92**, 2401–2435 (2012)
17. W. Tian, M.F. Chisholm, P.G. Khalifah, R. Jin, B.C. Sales, S.E. Nagler, D. Mandrus, Mater. Res. Bull. **39**, 1319–1328 (2004)
18. Y. Matsushita, H. Ueda, Y. Ueda, Nat. Mater. **4**, 845–850 (2005)
19. S. Das, X. Zong, A. Niazi, A. Ellern, J.Q. Yan, D.C. Johnston, Phys. Rev. B **76**, 054418 (2007)
20. E.M. Levin, C.R. Robbins, H.F. McMurdie, *Phase Diagrams for Ceramists* (The American Ceramic Society, Columbus, Ohio, 1964); subsequent volumes by different authors
21. M. Hansen, *Constitution of Binary Alloys* (McGraw-Hill, New York, 1958)
22. R.P. Elliot, *Constitution of Binary Alloys, First Supplement* (McGraw-Hill, New York, 1965)
23. F.A. Shunk, *Constitution of Binary Alloys, Second Supplement* (McGraw-Hill, New York, 1969)
24. W.G. Moffatt, *Handbook of Binary Phase Diagrams* (General Electric, New York, 1981)
25. T.B. Massalski, *Binary Alloy Phase Diagrams*, 2nd edn. (ASM International, Materials Park, OH, 1990)
26. H. Okamoto, *Desk Handbook: Phase Diagrams for Binary Alloys* (ASM International, Materials Park, OH, 2000)
27. R.E. Barks, D.M. Roy, in *Crystal Growth*, ed. by H.S. Peiser (Pergamon, Oxford, 1967), pp. 497–504
28. I. Shindo, H. Komatsu, Yogyo-Kyokai-Shi **85**, 22–26 (1977). (in Japanese)

Chapter 4
Choosing a Flux

As we saw in Chap. 3, phase diagrams provide valuable information for flux growth: the position and size of the liquidus curve set the appropriate growth conditions, which in turn determine the maximum yield of crystals. Phase diagrams also reveal which fluxes to avoid, such as when the formation of unwanted compounds or solid solutions is expected. One thing that is missing from phase diagrams, however, is information on the resulting crystals. Phase diagrams by themselves cannot be used to predict whether the crystals will be large or small, well-faceted or dendritic, and of high quality or containing many defects and inclusions. While these features depend on the combination of various factors, the limits of what can be achieved in size and quality are often determined by the properties of flux, and how the flux interacts with the solute at high temperatures.

A wide variety of elements, compounds, and combinations of compounds have been tried as a flux. However, a small number of fluxes have been used successfully on many occasions and it is useful to discuss their characteristics. In this chapter, we will first review the main requirements of a good flux, and then look at a number of well-known fluxes. Because oxides and intermetallic compounds have different sets of appropriate fluxes, these two groups of compounds will be discussed separately.

4.1 Properties of an Ideal Flux

Before we discuss the properties of specific fluxes, it is useful to think of an ideal flux. Because such a flux does not exist in most real cases, a compromise is usually made and an optimum flux is chosen on the basis of the most important requirements. Assuming that the goal is to grow large and high-quality crystals, an ideal flux should have the following properties:

M. Tachibana, *Beginner's Guide to Flux Crystal Growth*, NIMS Monographs,
DOI 10.1007/978-4-431-56587-1_4

1. A high solubility for the solute.
2. An appreciable change in solubility with temperature.
3. Will not form a compound or solid solution with the solute.
4. A low melting point.
5. A low volatility at growth temperatures.
6. A low viscosity at growth temperatures.
7. Will not react with the crucible.
8. Easily separated from grown crystals.
9. A low toxicity.
10. Available in a pure form at low cost.

We now examine each property in more detail.

The flux has a high solubility for the solute, an appreciable change in solubility with temperature, and does not form a compound or solid solution with the solute. In many cases, these three properties become the top priorities in choosing an appropriate flux. On the atomic scale, the dissolving power of flux comes from its ability to weaken the chemical bonds that exist between the solute ions. Because this dissolution process takes place with the formation of new bonds between the flux and the solute, the flux should have bonding properties similar to those of the solute.

For example, in the case of oxides, the main binding force that holds the constituent ions together is electrostatic Coulomb (ionic) interactions. To break the strong Coulomb interactions and keep the ions separated from each other in solution, the flux must also be ionic, and this limits the choice of flux to either oxides or halide salts. Moreover, the interactions between the flux and solute ions must not be too strong, as stronger interactions will lead to the formation of an unwanted compound—the term "solvation" is used to describe the formation of weak solute–flux complexes in the solution. In the case of intermetallic compounds, Coulomb interactions are much weaker and metallic elements are often used as a flux.

Frequently, a useful flux forms a stable compound with the solute at lower temperatures, or in a concentration range that does not interfere with the flux growth. For example, both PbO and B_2O_3 react with Al_2O_3 to form the compounds $PbAl_2O_4$ (incongruent melting at 980 °C) and $Al_4B_2O_9$ (incongruent melting at 1035 °C) [1]. These compounds are not crystallized when the PbO–B_2O_3 flux is present in a much larger quantity than Al_2O_3, and their existence partly explains why the flux has high dissolving power for many Al-based oxides.

The strength of solute–flux interactions also controls the temperature dependence of solubility. It is important to remember that it is the change in solubility with temperature, rather than the absolute solubility, that determines the yield of crystals on slow cooling. Although the evaporation method may be used when the solubility has weak temperature dependence, it works only when the flux is highly volatile and evaporation does not interfere with the growth experiment. For a slow-cooling growth, solubility must change by at least several percent if large crystals are to be grown in a reasonably sized crucible.

Although a good flux will be chemically similar to the solute, there must be some crystal-chemical differences to avoid solid solubility between the flux and the solute. A large difference in ionic radius is usually effective in limiting solid solubility, and this is why some of the most popular fluxes for oxides contain either very large ions (such as Pb^{2+} and Bi^{3+}) or very small ions (such as B^{3+}). A difference in ionic charge (valence state) is also effective, but the crystal can still incorporate such flux ions by forming charged defects (such as vacancy or interstitial ions) or incorporating other impurities to maintain charge neutrality. To eliminate or reduce the problem of solid solubility, a flux that has at least one ion in common with the crystal is often selected.

The flux has a low melting point. A low-melting-point flux usually allows the growth experiment to be performed at low temperatures. This is desirable from two perspectives. In terms of crystal quality, lower temperatures mean a lower density of structural defects and less contamination from the crucible. From a more practical aspect, growth at low temperatures may allow the use of cheaper and/or more convenient materials for the crucible and furnace protection. Since the practical aspects of flux growth will be discussed in Chap. 5, it is sufficient to point out here that the majority of fluxes for oxides have a melting point below 900 °C, and those for intermetallic compounds are mostly below 700 °C.

The flux has low volatility at growth temperatures. A volatile flux often requires the use of a sealed crucible, unless the evaporation method is being used to grow crystals. Significant evaporation of the flux often limits the quality of crystals, as it leads to uncontrolled nucleation and crystal growth at the liquid surface. On the other hand, a moderate evaporation loss can result in a larger yield of crystals, and it may even improve the quality of crystals in some cases.

The flux has low viscosity at growth temperatures. A flux with low viscosity is desirable, because it promotes diffusive atomic motion to achieve a homogeneous solution. As discussed in Chap. 2, the rate of stable growth is limited by the transport of solute and flux ions, and failure to achieve stable growth leads to the generation of defects such as flux inclusions. If the solution is highly viscous, simply lowering the cooling rate may not be adequate to achieve stable growth. It is often quoted that an ideal flux has viscosity in the range 1–10 cP (1 cP = 10^{-3} Pa·s), with a maximum practical viscosity of about 1000 cP [1].[1] For example, the glass-forming compound B_2O_3 is usually too viscous to be used as a flux by itself, so other compounds are added to reduce the viscosity. Compared with oxide systems, metallic fluxes and solutions are usually less viscous, so faster cooling rates are used frequently; 5 °C per hour, which is usually too fast for oxide growth, is often adequate for the growth of intermetallic compounds. It is important to keep in mind that viscosity rises rapidly with a decrease in temperature, and the addition of solute can greatly modify the viscosity.

---

[1]As a reference, the viscosities of water and honey at room temperature are about 1 and 5000 cP, respectively.

In the growth of some oxides, the addition of a small amount of B_2O_3 improves the size and quality of crystals. This has been attributed to the formation of complexes between B_2O_3 and solute ions [1], which increases both the viscosity and metastable region for crystallization. As this example shows, a moderate increase in viscosity can be beneficial in some cases.

The flux does not react with the crucible. A suitable flux must not react with the crucible. If a reaction occurs, the crucible material can end up in the growing crystals, and there may be leakage of hot solution in severe cases. Although the properties of crucible materials will be discussed in Chap. 5, it should be pointed out here that platinum crucibles are often used for the growth of oxides, silica glass for chalcogenides, and alumina or tantalum for many intermetallic compounds. These materials are chosen mainly on the basis of their stability towards the solution at high temperatures.

The flux can be easily separated from grown crystals. In order to collect the grown crystals from the crucible, there must be a way of removing the residual flux. This can be accomplished easily if water or some aqueous reagent dissolves the flux without attacking the crystals. Alkali halide fluxes are useful in this respect, as they are dissolved rapidly in water. In many cases, an acid, such as nitric acid (HNO_3) or hydrochloric acid (HCl), or an alkali solution, such as sodium hydroxide (NaOH), can be used to dissolve the flux.

If there is no appropriate solvent for this purpose, the residual flux may be removed while it is still molten at high temperatures. If the growth is carried out in an unsealed crucible, a pair of tongs can be used to pour off the liquid flux. If the crucible is sealed in a silica glass tube, the tube is inverted and placed in a centrifuge to separate the flux from the crystals. These techniques will be explained in Chap. 5.

The flux has low toxicity. Although most chemicals are toxic at some level, some chemicals are clearly more toxic than others. Some of the most highly toxic chemicals, which are classified as poisonous substances, include compounds of Be, As, Tl, and Hg. The chemicals classified as toxic substances, such as compounds of Pb and Cd, are less toxic but can still pose great health hazards. The crystal grower must follow the rules on the use of these substances (many research groups and institutions have their own rules), and be fully aware of proper handling procedures.

The flux is available in pure form at low cost. Websites of chemical suppliers show that different chemicals have widely different price tags, and that price is strongly dependent on purity. An appropriate choice of purity depends on the uses of the crystals, but 3 N (99.9%) or 4 N (99.99%) purity is often a good choice for flux materials—2 N is too impure in many cases, and 5 N and above are usually too expensive and the additional purity often does little to improve the crystal quality. (When semiconductor crystals are used for detailed transport studies, the starting chemicals should be of the highest purity.) Impure chemicals may be tolerated if the impurities are not incorporated into the crystals, but this is difficult to predict before the experiment.

As a summary of this section, we should remember that an optimum flux for oxides is often obtained by using a mixture of compounds, which has a number of advantages. (The same is true for intermetallic compounds, but it is much less

Fig. 4.1 Single crystals of $PrAlO_3$ obtained from $PbO–PbF_2–B_2O_3$ flux (right) and from the same flux with about 4 wt% MoO_3 added (left). Based on recipes originally reported in [2]

explored.) For example, a mixture can retain the good solvent properties of a volatile flux while the vapor pressure is being decreased. Mixtures can also allow reductions in viscosity and crystallization temperature, and they may improve the size and habit of the crystals. It should be emphasized that even an additive, at the level of a few percent, can dramatically affect the outcome in some cases (see Fig. 4.1 for an example), although the atomic mechanism is not understood in most such cases.

4.2 Typical Fluxes for Oxide Growth

Because there are so many possible candidates, it is often difficult to find a starting point when choosing a flux. A useful practice is to examine the literature for fluxes that have been used to grow similar compounds. Elwell and Scheel's book [1] provides comprehensive references up to 1975, and a review chapter by Wanklyn [3] has a wealth of examples for oxides and fluorides, also up to 1975. As an attempt to cover more recent examples, the Appendix of this book lists works published in the *Journal of Crystal Growth* since 1975. In this section, we discuss some of the major types of flux that are used in the growth of oxide crystals. Table 4.1 gives the properties of typical fluxes for oxides, and some of the crystals grown from these fluxes are shown in Table 4.2. In order to avoid a long list of references, only selected examples from the above sources [1, 3] are included in Table 4.2; the reader is directed to these sources for the original references.

Table 4.1 Properties of typical fluxes used for oxide growth

Flux	T_{melt} (°C)	T_{boil} (°C)[a]	Soluble in	Density[b]
PbO	886	1472	HNO_3	9.53
PbF_2	855	1293	HNO_3	8.445
$PbCl_2$	501	950	H_2O	5.85
Bi_2O_3	817	1890	HNO_3	8.90
BiF_3	727	1027	HNO_3	5.32
$PbO–B_2O_3$	>493	–	HNO_3	–
$PbO–PbF_2$	>495	–	HNO_3	–
$PbO–PbF_2–B_2O_3$	>494	–	HNO_3	–
$PbO–V_2O_5$	>480	–	HNO_3	–
$Bi_2O_3–V_2O_5$	>817	–	HNO_3	–
$Bi_2O_3–B_2O_3$	>622	–	HNO_3	–
B_2O_3	450	1860	H_2O	2.55
$LiBO_2$	849	–	HNO_3	2.223
$Li_2B_4O_7$	930	–	HNO_3	–
$NaBO_2$	966	1434	H_2O	2.46
$Na_2B_4O_7$	743	1575	H_2O	–
KBO_2	950	–	H_2O	–
$K_2B_4O_7$	815	–	H_2O	–
$BaO–B_2O_3$	>740	–	HNO_3	
V_2O_5	690	1750	Acids	3.357
$LiVO_3$	616	1750	H_2O	–
$NaVO_3$	630	–	H_2O	–
MoO_3	795	1155	Acids	4.69
$Li_2O–MoO_3$	>532	–	Alkali	–
$Na_2O–MoO_3$	>499	–	Alkali	–
$K_2O–MoO_3$	>480	–	Alkali	–
WO_3	1473	1700	Alkali	7.16
$Li_2O–WO_3$	>695	–	Alkali	–
$Na_2O–WO_3$	>480	–	Alkali	–
$K_2O–WO_3$	>550	–	Alkali	–
LiF	845	1676	H_2O	2.635
LiCl	605	1382	H_2O	2.068
NaF	993	1704	H_2O	2.558
NaCl	801	1413	H_2O	2.165
KF	858	1502	H_2O	2.48
KCl	770	1420	H_2O	1.984
$CaCl_2$	772	1635	H_2O	2.15
$SrCl_2$	874	1250	H_2O	3.052
$BaCl_2$	962	1560	H_2O	3.856
$CdCl_2$	568	964	H_2O	4.047

(continued)

Table 4.1 (continued)

Flux	T_{melt} (°C)	T_{boil} (°C)[a]	Soluble in	Density[b]
Li_2CO_3	723	1310	H_2O	2.11
Na_2CO_3	851	1600	H_2O	2.54
K_2CO_3	891	–	H_2O	2.43
NaOH	318	1388	H_2O	2.13
KOH	360	1327	H_2O	2.12

[a]Decomposition temperature in some cases, [b]near room temperature, in g/cm^3

The fluxes in the tables can be classified into five major types:

1. Lead- and bismuth-based polar compounds.
2. Network-forming borates.
3. Complex-forming vanadates, molybdates, and tungstates.
4. Simple ionic alkali halides and carbonates.
5. Oxidizing alkali hydroxides.

4.2.1 Lead- and Bismuth-Based Polar Compounds

PbO, PbF_2, and, to a lesser extent, Bi_2O_3, have been used to grow single crystals of many oxides. They have high solubility for many types of oxides, which is attributed to their large polarizabilities (this is similar to why polar H_2O is a good solvent for many salts). Furthermore, the crystals grown from these fluxes tend to be large and of high quality, which have been attributed to the formation of favorable complexes between solute and flux in the solution. The three compounds have quite distinct melting and boiling points, and their properties as a flux can be improved by mixing together these compounds and/or adding B_2O_3. These fluxes often do not dissolve oxides containing alkali metal ions, however.

PbO is not usually used as a flux by itself. This is because its volatility can be suppressed by adding B_2O_3, and the solubility is often increased by adding PbF_2 or B_2O_3. These additions can also reduce attacks on the platinum crucible, which occur when PbO becomes Pb under a reducing condition. (Replacing some of the PbO by PbO_2 or Pb_3O_4 can help to maintain an oxidizing environment [3].) PbO–B_2O_3, PbO–PbF_2, and PbO–PbF_2–B_2O_3 are some of the most popular fluxes for oxides, with typical growth temperatures being 1250–900 °C.

PbF_2 has high vapor pressures at these temperatures, and it is the first component to evaporate from mixed flux systems. In some cases, the change in the flux composition leads to inhomogeneities in the crystals. The evaporation loss can be prevented by tightly crimping a lid onto the platinum crucible, but if PbF_2 alone is used as a flux, perfect sealing by welding is usually necessary. On the other hand, the high volatility of PbF_2 can be utilized to grow crystals by the evaporation technique. It must be remembered, however, that the evaporated fume is toxic and

Table 4.2 Examples of oxide flux growth, known before 1975 (*see* the Appendix for more recent examples)

Flux	Crystal[a]
PbO	$PbFe_{12}O_{19}$, $PbFe_{1/2}Ta_{1/2}O_3$, $PbFe_{1/2}Nb_{1/2}O_3$, $PbMg_{1/3}Nb_{2/3}O_3$, $PbSc_{1/2}Nb_{1/2}O_3$, $RAlO_3$, $RFeO_3$, $R_3Fe_5O_{12}$, $LaZn_{1/2}Ru_{1/2}O_3$, $LiFe_5O_8$, Mg_2SiO_4, $CaSiO_3$, $CaMg(SiO_3)_2$, $NiFe_2O_4$, $ZnFe_2O_4$, $ZnRh_2O_4$, R_2GeO_5, $R_2Ge_2O_7$
PbF_2	$PbTiO_3$, Al_2O_3, MnO, ZnO, NiO, Ga_2O_3, In_2O_3, SnO_2, ZrO_2, CeO_2, $SrTiO_3$, $LaNbO_4$, $Bi_2Sn_2O_7$, $PbZrO_3$, $Pb_3Ta_4O_{13}$, MgO, BeO, TiO_2, HfO_2, ThO_2, NpO_2, $RCrO_3$, $RMnO_3$, $SrTiO_3$, $MgAl_2O_4$, $MgFe_2O_4$
Bi_2O_3	$Bi_4Ti_3O_{12}$, $(Ba,Bi)FeO_3$, $Bi_2Fe_4O_9$, Fe_2O_3, $RCrO_3$, $RMnO_3$, R_2GeO_5, R_2SiO_5, $R_2Ge_2O_7$
$PbO–B_2O_3$	$PbTiO_3$, Al_2O_3, BeO, Ga_2O_3, Fe_2O_3, In_2O_3, $RFeO_3$, $Y_3Fe_5O_{12}$, $LiFe_5O_8$, $LiGaO_2$, RBO_3, $LiAlO_2$, $MgFe_2O_4$, $NiFe_2O_4$
$PbO–PbF_2$	$PbFe_{12}O_{19}$, $PbSc_{1/2}Ta_{1/2}O_3$, $PbMn_2O_4$, Al_2O_3, Fe_2O_3, Mn_3O_4, ZnO, CeO_2, $RAlO_3$, $R_3Al_5O_{12}$, $LaNi_{1/2}Ru_{1/2}O_3$, $LaNi_{1/2}Ir_{1/2}O_3$, $LaMg_{1/2}Ru_{1/2}O_3$, $RFeO_3$, $R_3Fe_5O_{12}$, $R_3Ga_5O_{12}$, $RMnO_3$, $(La,Pb)MnO_3$, $BeAl_2O_4$, $CoFe_2O_4$, $CuFe_2O_4$, $MgGa_2O_4$, $ZnGa_2O_4$, RMn_2O_5, R_3NbO_7, ROF, $R_2Ti_2O_7$, $NiFe_2O_4$, $FeBO_3$, $(Cr, Mn)_3O_4$
$PbO–B_2O_3–PbF_2$	$MgAl_2O_4$, $Y_3Fe_5O_{12}$
$PbO–V_2O_5$	RVO_4, TiO_2, Fe_2O_3, Ga_2O_3, ThO_2, Fe_2TiO_5, $NiTiO_3$, $GaFeO_3$, $PbTa_2O_6$
$Bi_2O_3–V_2O_5$	RVO_4, Al_2O_3, Ga_2O_3, Cr_2O_3, Fe_2O_3, $R_3Al_5O_{12}$, $R_3Fe_5O_{12}$, $NiFe_2O_4$, $RNbO_4$
$Li_2O–B_2O_3$	Cr_2O_3, Fe_2O_3
$Na_2O–B_2O_3$	ThO_2, BeO_2, Al_2O_3, TiO_2, NiO, Mn_3O_4, Fe_2O_3, Fe_3O_4, ZrO_2, HfO_2, CeO_2, UO_2, MgO, Cr_2O_3, In_2O_3, $(Zn,Sb)_3O_4$, $ThTi_2O_6$
$BaO–B_2O_3$	$BaNb_2O6$, $BaZn_2(Fe,Al)_{12}O_{22}$, $BaTa_2O_6$, $BaTiO_3$, $Ba(Lu,Ta)O_3$, NiO, $Y_3Fe_5O_{12}$, $Y_3Al_5O_{12}$, $RFeO_3$, $NiFe_2O_4$
V_2O_5	$LiAlSiO_4$, VO_2, V_2O_3, $FeVO_4$
$Li_2O–V_2O_5$	$ZrSiO_4$
$Na_2O–V_2O_5$	RVO_4, Fe_2O_3
$Li_2O–MoO_3$	$Li_2M_2(MoO_4)_3$ ($M =$ Co, Ni, Cu), BeO, TiO_2, SiO_2, GeO_2, MoO_3, CeO_2, WO_3, ThO_2, NpO_2, $CaWO_4$, Al_2SiO_5, $CaTiSiO_5$, Zn_2SiO_4, $ZrSiO_4$, R_2SiO_5, $HfSiO_4$, $ThSiO_4$, $ZrTiO_4$, Be_2SiO_4
$Li_2O–WO_3$	$LiR(WO_4)_2$, CeO_2, GeO_2, ThO_2, BeO, $ZrSiO_4$, $HfSiO_4$, $ThSiO_4$
$Na_2O–MoO_3$	$(Cr,Mn)_3O_4$, $Al_2(WO_4)O_3$
LiCl	$LiMPO_4$ ($M =$ Fe, Mn, Co, Ni)
NaCl	TiO_2, CaO, CoO, NiO, CuO, $CaSiO_3$, $CaMoO_4$, $CaWO_4$, $SrSO_4$, $BaSO_4$
KF	$K_2Fe_{12}O_{19}$, $K_2Ga_{12}O_{19}$, $K_2Ge_4O_9$, CeO, $PbTiO_3$, $CdTiO_3$, $CeAlO_3$, $SrTiO_3$, $BaTiO_3$, $Y_4Si_3O_{12}$, $R_2Si_2O_7$, $R_2O(SiO_4)$, $Bi_2Sn_2O_7$
$CaCl_2$	$Ca_5(PO_4)_3Cl$, $CaMn_2O_4$, $Ca_2Nb_2O_7$, $CaCr_2O_4$, $CaRuO_3$
$SrCl_2$	Sr_2NiWO_6, $SrRuO_3$
$BaCl_2$	$BaTiO_3$, $BaFeO_{19}$, $BaWO_4$, BaB_2O_4, $BaTi_3O_7$
Na_2CO_3	$NaNbO_3$, $NaLaFe_{12}O_{19}$, $BaFe_{12}O_{19}$
K_2CO_3	$KNbO_3$, $K_4Nb_6O_{17}$, $KTaO_3$
KOH	$KNbO_3$, MgO

[a]*R* = rare earth, although not necessarily meaning all members of the rare earth

corrodes many materials. Both PbO and PbF_2 are usually unsuited for the growth of oxides containing Ba^{2+}, Sr^{2+}, or Ca^{2+}, as Pb^{2+} shares the same charge and similar ionic radius with these ions.

Bi_2O_3 is less volatile than Pb-based flux, but it usually has less dissolving power. It is reduced easily to Bi metal, so the oxidizing condition must be maintained to prevent attack on the platinum crucible. One way of preventing reduction is to add V_2O_5 to the flux [3]: if reduction starts to occur, V_2O_5 will first change to V_2O_4, giving up its oxygen to protect Bi_2O_3. Although Bi_2O_3 has good dissolving power for rare-earth ions, the similarity in charge and size between Bi^{3+} and large trivalent rare-earth ions (such as La^{3+}) makes solid solubility almost inevitable. The problem lessens for the smaller rare-earth ions, and high-quality single crystals of $YMnO_3$ have been grown from Bi_2O_3 flux [4].

4.2.2 Network-Forming Borates

Although B_2O_3 has already been mentioned as a useful additive to Pb-based flux, it can also be used as a main flux component. B_2O_3 by itself is rarely used as a flux because of its high viscosity, which is due to the formation of a glassy network in the liquid state. The viscosity can be reduced significantly by adding alkali metal ions, which break the network of B–O bonds. $Na_2B_4O_7$ and $Li_2B_4O_7$ in particular have been used to grow many oxides, although the resulting crystals tend to be small. These compounds can be purchased as chemical reagents, or they can be prepared by reacting Na_2CO_3/Li_2CO_3 with B_2O_3.

Because B^{3+} and alkali metal ions are not easily reduced to their metals, these fluxes have been used to grow oxides with transition metal ions in reduced oxidation states. Examples include: (1) LiV_2O_4, the platinum crucible is sealed in a silica glass tube [5]; (2) Fe_3O_4, grown in 2 atm of argon with 1 ppm of oxygen present [6]; and (3) MoO_2, grown in a nitrogen atmosphere at the oxygen partial pressure of 10^{-8} atm [7].

Another useful additive for B_2O_3 is BaO, which results in a nearly non-volatile flux. However, it has high viscosity, and excessive nucleation cannot be avoided without the use of a seed. For these reasons, BaO–B_2O_3 has mostly been used for top-seeded solution growth [1].

4.2.3 Complex-Forming Vanadates, Molybdates, and Tungstates

V_2O_5, MoO_3, and WO_3 are also used in the flux growth of many oxides. Because these oxides are volatile at high temperatures, alkali metal ions are often added to reduce volatility; the examples are A_2MO_4 and $A_2M_2O_7$ (A = Li, Na, or K; M = Mo or W). The solubility usually decreases with the addition of alkali metal ions. These fluxes are especially useful for the growth of silicates (compounds containing silicon and oxygen), which explains their frequent appearance in mineralogical works. V_2O_5 is also used as an additive to molybdate and tungstate fluxes.

In a classic study [8], the Lewis concept of acid and base was used to explain solvent action. At high temperatures, the $Na_2W_2O_7$ flux dissociates into a Lewis acid, WO_3, and a Lewis base, Na_2WO_4. (The Lewis acid is defined as an electron pair acceptor and the Lewis base is an electron pair donor.) When $Na_2W_2O_7$ is mixed with a solute and heated, the solute reacts with the liberated WO_3 to form a complex in the solution. On cooling, the complex reacts with Na_2WO_4, freeing up the solute for crystallization.

4.2.4 Simple Ionic Alkali Halides and Carbonates

Because of their low dissolving power for most oxides, in many cases alkali halides are used only when other fluxes fail for various reasons. KF is probably used most often within this group, and this flux has recently been used to grow $Pr_2Ir_2O_7$ and $Eu_2Ir_2O_7$ crystals [9]. Because the halides are volatile at high temperatures, a mixture of two eutectic alkali halides is frequently used to lower growth temperatures. Alumina crucibles are often used for chloride flux, as it tends to attack platinum at high temperatures.

Alkali carbonates have mostly been used as a self-flux (e.g., K_2CO_3 to grow $KNbO_3$ and $KTaO_3$ [10]), but they have recently been used to grow complex oxides containing platinum group metals [11].

The alkali-earth halides share similar properties with alkali halides. The chlorides $BaCl_2$, $SrCl_2$, and $CaCl_2$ have been used to grow complex oxides sharing the same alkali-earth metal. Examples are $SrRuO_3$, $CaRuO_3$ [12], and Sr_2NiWO_6 [13].

4.2.5 Oxidizing Alkali Hydroxides

In many ways, the hydroxides are very different from the other fluxes discussed so far. The melting points of hydroxides are very low (318 °C for NaOH and 360 °C for KOH), and some eutectics have melting temperatures below 200 °C. If the hydroxide (usually in the form of pellets) is left in open air, it will absorb moisture and eventually turn into an aqueous solution. When molten, the hydroxide ion is in equilibrium with water and O^{2-}, that is, $2OH^- = H_2O + O^{2-}$. The O^{2-} ion will react with oxygen in air to form $O_2{}^{2-}$ and $O_2{}^-$, which are highly oxidizing. Therefore, hydroxide fluxes can stabilize transition metals in high oxidation states, which otherwise require very high oxygen pressures. The growths of Ba_2NaOsO_6 and $Ba_3Mn_2O_8$ have been accomplished by this technique [14].

Alumina crucibles are used for growth in open air. The evaporation of hydroxide can be regulated by partial sealing, which controls both the growth rate and oxidizing environment. If a closed system is desired, a silver tube is used as the container (it can be easily crimped shut and flame sealed). In addition to alkali hydroxides, alkali-earth hydroxides are also effective with similar techniques. The chemistry and application of hydroxide fluxes have been reviewed recently [11, 15].

4.2.6 *Other Considerations*

The fluxes described above are the most typical ones for oxide growth. The above classification is not strict, and those belonging to different groups can be combined to produce a better flux. If one of the typical fluxes is a constituent of the target compound, or if there is a common metal between a typical flux and the target compound, that flux should be considered as a strong candidate. In some cases, a self-flux that is not in any of the above groups is used; we have seen in previous chapters that CuO is a useful self-flux for the growth of La_2CuO_4 and $CuGeO_3$, and TiO_2 is used for the growth of $BaTiO_3$ in the top-seeded solution growth technique. Similarly, if the target compound contains a constituent oxide that has a low melting point, that oxide may be included in the flux. When a mixture of two or more compounds is used as a flux, the eutectic composition can be a good choice, provided that it does not have high viscosity or high volatility. In some systems, the acid–base concept can be used to help assess the optimum composition of the flux [16–18]. (The acidic fluxes include B_2O_3, V_2O_5, MoO_3, and WO_3, whereas the basic fluxes include PbO, Bi_2O_3, and alkali oxides.)

4.3 Fluxes for Intermetallic Compounds

Intermetallic compounds have different chemical bonding characteristics from oxides. Also, these compounds react with oxygen at high temperatures, such that crystal growth must be carried out in an oxygen-free environment. For these reasons, most of the fluxes used to grow oxide crystals are not suitable for the growth of intermetallic compounds, with the exception of chloride fluxes that are used in some cases.

Table 4.3 shows the properties of common fluxes used for intermetallic compounds, and Table 4.4 provides examples of compounds that have been grown from these fluxes. (Again, to avoid a long list of references, only examples from [1, 19, 20] are included in Table 4.4.) It is immediately obvious from these tables that elements of simple metals are widely used. Also, unlike the case of oxides, combinations of elements are not often used as flux for intermetallic compounds. The flux elements shown have a low melting point, and are relatively stable in air at room temperature. If the target compound contains one of these elements as a

Table 4.3 Properties of typical metallic fluxes used for the growth of intermetallic compounds

Flux	T_{melt} (°C)	T_{boil} (°C)	Soluble in	Density[a]
Al	660	2470	HCl	2.70
Cu	1085	2562	HNO_3	8.96
Zn	420	907	HCl	7.14
Ga	30	2400	HNO_3	5.91
Cd	321	767	HNO_3	8.65
In	157	2072	HNO_3	7.31
Sn	232	2602	HCl	7.265
Sb	631	1635	HNO_3	6.697
Hg	−39	357	HNO_3	13.534
Pb	327	1749	HNO_3	11.34
Bi	272	1564	HNO_3	9.78

[a]Near room temperature, in g/cm^3

Table 4.4 Examples of crystal growth from metallic flux

Flux	Crystal[a]
Al	TiB_2, ZrB_2, $RAlB_4$, RB_4, RB_6, RBe_{13}, UBe_{13}, Si, Ge, RAl_3, AlAs, AlB_2, Al_3Er, AlP, AlSb, $MnAl_6$, $NiAl_3$
Cu	RRh_4B_4, RCu_2Si_2, V_3Si, RIr_2, UIr_3, MnSi
Zn	InSb, GaSb, InAs, Si, Ge, ZnSe, ZnTe
Ga	Si, Ge, GaSb, RSb, $R_2Pt_4Ga_8$, CdS, GaAs, GaP, GaSb, Ga_2Se_3, $MnGa_6$, VGa_5, ZnS, ZnSe, ZnTe
Cd	CdS, CdSe, $CdSnP_2$, CdTe, Ge
In	RCu_2Si_2, Si, Ge, RIn_3, RCu_2Ge_2, RNi_2Ge_2, CdS, $CuGa_{1-x}In_xS_2$, InAs, InP, InS, InSb, ZnS, ZnSe, ZnTe
Sn	RCu_2Si_2, $R_3Rh_4Sn_{13}$, RFe_4P_{12}, $ZnSnP_2$, GaSb, GaP, $ZnSiP_2$, $CdSiP_2$, RSn_3, TiNiSn, MnSnNi, RSb, $CdGeP_2$, CdS, CdSe, $CdSiAs_2$, $CdSnP_2$, CdTe, CuP_2, Ge, $MgGeP_2$, Nb_3Sn, Si, $SnZnAs_2$, $ZnGeP_2$, ZnS, ZnSe, $ZnSnAs_2$, $ZnSnP_2$, $ZnSnSb_2$, ZnTe, $CdSnP_2$, $ZnSnSb_2$
Sb	$ZnSiP_2$, $CdSiP_2$, RSi_2, RSb_2, $U_3Sb_4Pt_3$, $PtSb_2$, GaSb, Ge, ZnSb
Hg	InSb, HgS, HgSe, HgTe, Mn_5Ge_3, Mn_5Si_3, Ni_2Si, Pt_2Si, RGe_2, RN, RSi_2
Pb	RPt_2, GaSb, RPb_3, RPbPt
Bi	UPt_3, PtMnSb, NiMnSb, UAl_3, UIr_3, GaP, $ZnSiP_2$, $CdSiP_2$, YPd, RBiPr, $R_3Bi_4Pt_3$, RBi_2, CdS, CdSe, CdTe, $ZnGeP_2$, ZnS, ZnSe, ZnTe

[a]*R* = rare earth, although not necessarily meaning all members of the rare earth

constituent, it is often a prime candidate as a self-flux. It is usually difficult to predict which flux will produce the best result; however, the same flux often works well for different compounds in the same structural group. It is also important to examine the available phase diagrams of the flux and solute combination. If there are other stable compounds in the relevant part of the phase diagram, these can interfere with the growth of the desired compound.

If none of the common fluxes produces the desired crystals, it is useful to check whether there is a low-melting-point composition within the phase field of constituent elements. For example, the best flux for the growth of RNi_2B_2C (R = rare earth) was found to be Ni_2B [21], which has a relatively low melting point of 1125 °C. The use of FeAs (melting point of 1070 °C) as a flux for the growth of $BaFe_2As_2$-type compounds is another good example.

An important factor to consider is the reactivity of the solution (solute and flux) with the container material. Because growth must be carried out in an oxygen-free environment, the solution is often sealed in a container. If the solution does not react with silica (SiO_2), as in the case of many chalcogenides with a halide flux, a sealed silica glass tube can be used as a crucible. If the solution reacts with silica but not with Al_2O_3, an alumina crucible can be sealed in a silica glass tube. In some cases, the solute becomes less reactive when it is diluted with the flux. For example, although a pure melt of rare-earth metals reacts with Al_2O_3, about ten atomic percent of rare earth in low-melting-point metals do not attack the alumina crucible up to 1200 °C [22]. In case alumina is not a viable material for the crucible, other materials such as Ta, BN, CaO, Y_2O_3, and graphite must be considered.

References

1. D. Elwell, H.J. Scheel, *Crystal Growth from High-Temperature Solutions* (Academic Press, London, 1975)
2. B.M. Wanklyn, S.H. Smith, G. Garton, J. Cryst. Growth **33**, 150–154 (1976)
3. B.M. Wanklyn, in *Crystal Growth*, ed. by B.R. Pamplin (Pergamon, Oxford, 1975), pp. 217–288
4. M. Tachibana, J. Yamazaki, H. Kawaji, T. Atake, Phys. Rev. B **72**, 064434 (2005)
5. Y. Matsushita, H. Ueda, Y. Ueda, Nat. Mater. **4**, 845–850 (2005)
6. M. Vichr, H. Makram, J. Cryst. Growth **5**, 77–78 (1969)
7. T. Sekiya, Mater. Res. Bull. **16**, 841–846 (1981)
8. W. Kunnmann, A. Ferretti, R.J. Arnott, D.B. Rogers, J. Phys. Chem. Solids **26**, 311–314 (1965)
9. J.N. Millican, R.T. Macaluso, S. Nakatsuji, Y. Machida, Y. Maeno, J.Y. Chan, Mater. Res. Bull. **42**, 928–934 (2007)
10. D. Rytz, H.J. Scheel, J. Cryst. Growth **59**, 468–484 (1982)
11. D.E. Bugaris, H.-C. Zur Loye, Angew. Chem. Int. Ed. **51**, 3780–3811 (2012)
12. R.J. Bouchard, J.L. Gillson, Mater. Res. Bull. **7**, 873–878 (1972)
13. C.G.F. Blum, A. Holcombe, M. Gellesch, M.I. Sturza, S. Rodan, R. Morrow, A. Maljuk, P. Woodward, P. Morris, A.U.B. Wolter, B. Büchner, S. Wurmehl, J. Cryst. Growth **421**, 39–44 (2015)
14. I.R. Fisher, M.C. Shapiro, J.G. Analytis, Philos. Mag. **92**, 2401–2435 (2012)
15. S.J. Mugavero, W.R. Gemmill, P. Roof, H.-C. Zur Loye, J. Solid State Chem. **182**, 1950–1963 (2009)
16. B.M. Wankly, J. Cryst. Growth **37**, 334–342 (1977)
17. B.M. Wankly, J. Cryst. Growth **43**, 336–344 (1978)
18. B.M. Wankly, J. Cryst. Growth **65**, 533–540 (1983)

19. Z. Fisk, J.P. Remeika, in *Handbook on the Physics and Chemistry of Rare Earths*, vol. 12, ed. by K.A. Gschneider Jr., L. Eyring (Elsevier, Amsterdam, 1989), pp. 53–70
20. P.C. Canfield, Z. Fisk, Philos. Mag. B **65**, 1117–1123 (1992)
21. P.C. Canfield, I.R. Fisher, J. Cryst. Growth **225**, 155–161 (2001)
22. P.C. Canfield, in *Properties and Applications of Complex Intermetallics*, ed. by E. Belin-Ferré (World Scientific, Singapore, 2010), pp. 93–111

Chapter 5
Equipment and Experimental Procedures

We looked briefly at a flux growth experiment in Sect. 1.2, but now is the time to look at the equipment and procedures in more detail. In particular, we look at various types of furnaces and crucibles, and the starting materials used in the experiments. It will become clear that the types of equipment necessary are quite simple and, unlike some other techniques of crystal growth, can be readily obtained commercially. The second half of this chapter describes the basic procedures of flux growth experiments, one section covering growth in air, and another, growth in a protective atmosphere. The final section provides some suggestions on what to do after the crystals are obtained.

It must be emphasized that there are potential dangers and health hazards in carrying out flux growth. When working with hot furnaces or open flames, eyes must be protected by a face shield or safety glasses, and hands by thermally insulating gloves. The handling of most powdered chemicals requires the use of an appropriate mask and plastic gloves. The starting point in preventing any accident is to evaluate and anticipate potential hazards before the experiment; anyone who is unable to do so should not attempt the experiment.

5.1 Furnaces

Most experiments of flux growth take place within the temperature range 700–1300 °C. Such temperatures can be conveniently attained in electric resistance furnaces, or resistance furnaces for short. When electricity is passed through a resistive material, heat is produced at a rate proportional to the resistance and to the square of the current passed through the material. By using resistive conductors called heating elements, and closely regulating the amount of electrical power, precise temperature control can be achieved with relative ease.

Resistance furnaces are available in a wide range of sizes, designs, and temperature capabilities, and at an equally large range of prices. A wrong choice of

M. Tachibana, *Beginner's Guide to Flux Crystal Growth*, NIMS Monographs,
DOI 10.1007/978-4-431-56587-1_5

furnace can severely limit its usefulness in flux growth, or markedly increase the operation costs. To help make an appropriate choice, we discuss some considerations that go into the design and materials used in resistance furnaces.

The operating principle of a resistance furnace is shown in Fig. 5.1. For use in flux growth, the temperature controller must be programmable and of high quality—it must be possible to change the temperature at a controlled and stable rate. This is because fluctuations in temperature during the cooling period can lead to spurious nucleation and unstable crystal growth. In general, the temperature controller must employ proportional control and the precision should be better than ±0.3 °C. Many low-end controllers use a simple on–off operation and do not satisfy this standard.

There are two types of furnaces that are particularly useful for flux growth: the vertical tube furnace (Fig. 5.2a) and the box furnace (Fig. 5.2b). Here, we discuss the features of these furnaces with mostly oxide growth in mind, but other types of experiments can be carried out with little or no modifications.

5.1.1 *Vertical Tube Furnace*

The vertical tube furnace uses two ceramic tubes of different diameters in the upright position: the outer tube separates the crucible space from the heating element, while the inner tube holds the crucible in an appropriate position. The heating element may be in the form of a tube, or made up of rows of rods or coiled wires around the outer ceramic tube. One advantage of the vertical tube furnace is the presence of the outer ceramic tube, which helps to protect the heating element from corrosive flux. If flux vapor escapes from the crucible, it will mostly rise up along the tube and condense at the cooler top, rather than penetrate through the tube at the

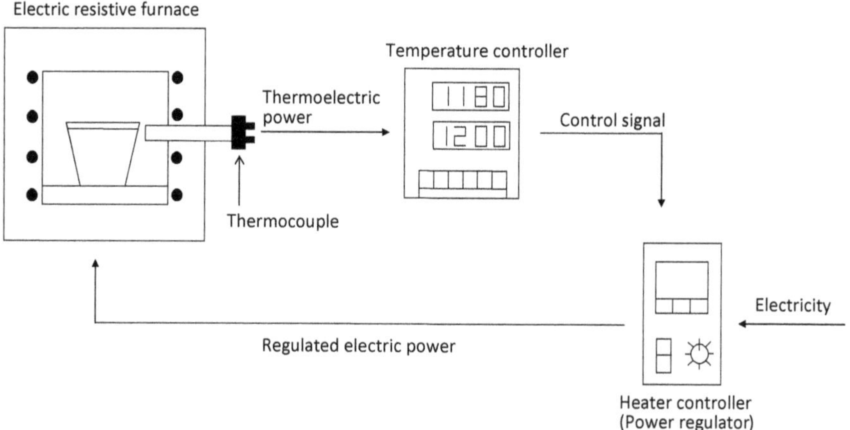

Fig. 5.1 Operating principle of an electric resistance furnace

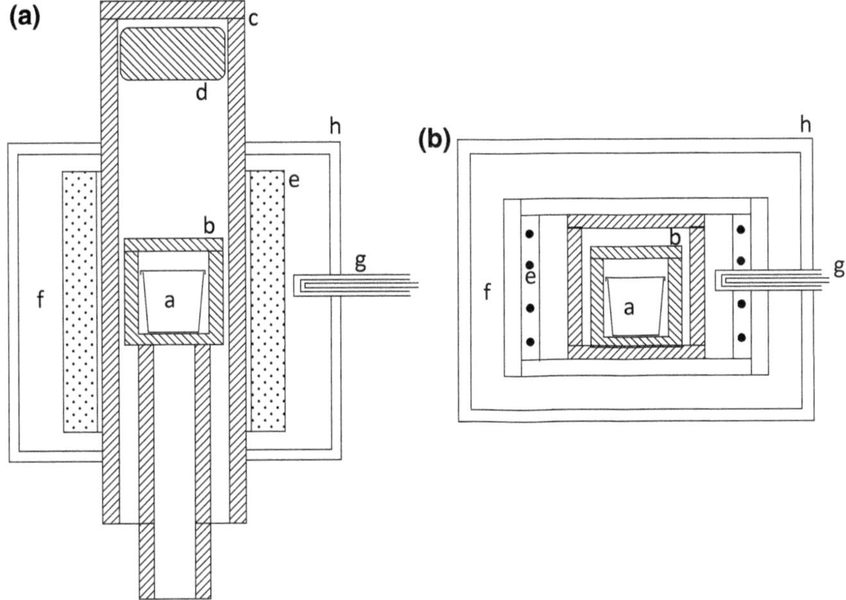

Fig. 5.2 **a** Box furnace and **b** vertical tube furnace. a: Platinum crucible; b: alumina crucible; c: ceramic tube; d: silica wool; e: heating element; f: ceramic insulation; g: thermocouple; h: metal casing

hottest central region. The damaged ceramic tube (*see* Fig. 5.3) can be replaced at low cost, leaving no permanent damage to the rest of the furnace.

The two most common materials used for ceramic tubes are sintered alumina and mullite. Alumina tubes are made of 95–99% pure Al_2O_3, with the rest being mostly SiO_2. Mullite tubes typically have a composition with the molar ratio of 60% Al_2O_3 and 40% SiO_2. Mullite tubes are more susceptible to damage from corrosive flux, but they cost only a quarter or less of their alumina counterparts. Alumina tubes have a higher maximum working temperature of 1800 °C, compared with 1600 °C for mullite tubes. Both alumina and mullite tubes are susceptible to thermal shock, meaning that they can crack if subjected to a sudden change in temperature.

As these ceramic tubes remain fairly airtight up to high temperatures, it is possible to introduce various gases into them. The gas flow is achieved by fixing sealing caps at both ends of the tube, and introducing appropriate gases from one end and removing them at the other end. With flowing argon gas, for example, air-sensitive materials such as intermetallic compounds can be grown in a tube furnace. The review article by Fisk and Remeika [1] describes this technique in some detail.

The vertical tube furnace also allows easy adjustment of the crucible position, providing access to various temperature gradients. When the heater power is turned

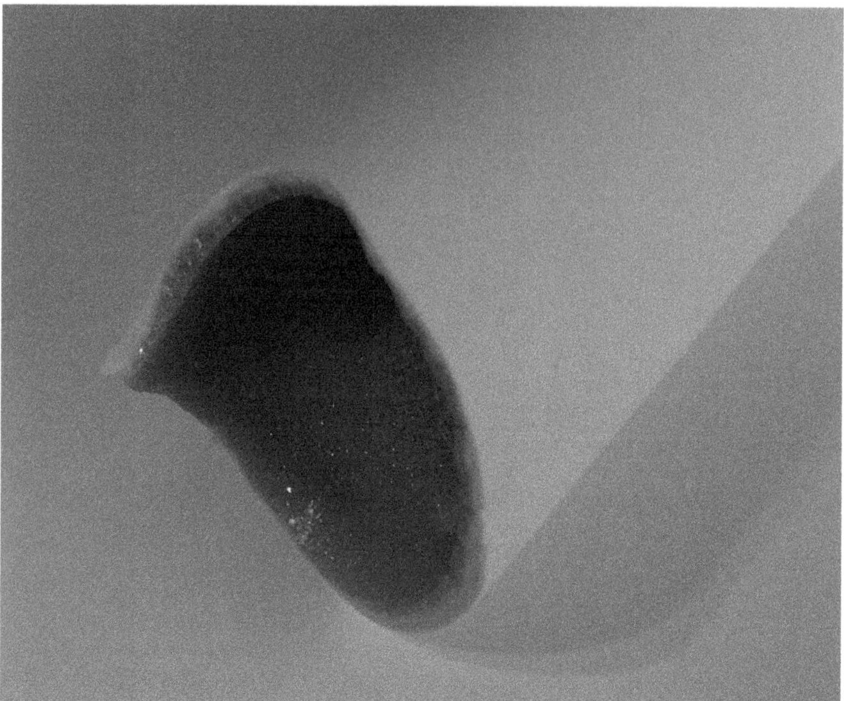

Fig. 5.3 Corroded mullite ceramic tube

on, the highest temperature is found at the center region. This is where the crucible is usually positioned, and it gives fairly uniform temperature along the height of the crucible. On the other hand, it is sometimes necessary to suppress nucleation at the surface of the solution, because crystals grown at the surface tend to have many defects. For this purpose, the crucible can be lowered from the center, which makes the bottom cooler and encourages nucleation at this position. Of course, the vertical temperature distribution must be known in advance to make the adjustment reliable. A cooler crucible bottom can also be obtained by sending an air jet through the inner ceramic tube, using a fish-tank pump.

A downside of the vertical tube furnace is that only one crucible can be used in a single experiment (although several ceramic crucibles may be stacked on top of each other, they will experience different temperature gradients). This limits the usefulness of vertical tube furnaces in exploratory works, where different compositions have to be tried in many experiments. Another drawback is the difficulty in removing the crucible at high temperatures. Such an operation is necessary when there is a need to decant, or pour off, the liquid flux at the end of the growth run, a process which often must be carried out in a matter of seconds.

5.1.2 *Box Furnace*

The two problems encountered in the use of vertical tube furnaces—only one crucible per run and difficulty in decanting—are solved when box furnaces are used instead. In the box furnace, which is also called a rectangular muffle furnace, heating elements are usually placed in the side walls and/or floor panel. A large door on the front side gives easy access to the inside of the furnace. Because the center region is designed to give a fairly uniform temperature distribution, several crucibles can be placed in a single run. By turning off the power and limiting the duration for which the door is open, damage to the heating elements can be minimized during the decanting process.

Although some box furnaces have a small hole on the ceiling, it does not provide an adequate escape route for flux vapors. Therefore, crucibles should be sealed in ceramic containers to minimize damage to the furnace. If heating elements are exposed inside the furnace, these areas should be covered with ceramic plates. Even with these precautions, flux vapors can damage the ceramic walls, heating elements, and thermocouples, so the furnace must be easy to repair.

As corrosive vapor can escape from both vertical tube and box furnaces, they must be placed in a well-ventilated area, preferably below a fume hood. Furnaces must also be free from vibration, as mechanical shock leads to spurious nucleation and growth instabilities. Anti-vibration supports can prevent small table-top furnaces from wobbling.

5.1.3 *Heating Elements*

The maximum usable temperature of a resistance furnace is normally determined by the heating element. Table 5.1 lists typical materials used for heating elements along with their maximum working temperatures. In most cases, it is much more convenient to use heating elements that can be used in open air; although materials such as molybdenum, tungsten, and carbon can be heated to high temperatures, these must be enclosed in a dynamic vacuum or an inert atmosphere. Although platinum-based heating elements can also be used in air, they are rarely used owing to their high cost. The temperatures quoted are only approximate and refer to the value at the crucible, not at the heating element, which may be up to 200 °C higher.

Ni–Cr (nichrome) heating elements are made of an alloy of mostly nickel and chromium. They are also used in oven toasters and hair dryers, and are available at low cost in the form of wires, ribbons, and strips. When they are heated in air for the first time, a layer of Cr_2O_3 is formed at the surface to prevent further oxidation. The low maximum working temperature (about 1000 °C) limits their usage in flux growth, but nichrome furnaces can be used for various auxiliary experiments, such as the premelting and cleaning of platinum crucibles.

Table 5.1 Properties of materials used for heating elements

Material	T_{max} (°C)[a]	T_{melt} (°C)	Use in air
Ni–Cr (nichrome)	1000	1400	Yes
Fe–Cr–Al (kanthal)	1200	1500	Yes
SiC (silicon carbide)	1400	2730	Yes
$MoSi_2$ (kanthal super)	1700	2030	Yes
$LaCrO_3$ (lanthanum chromite)	1800	2450	Yes
Pt (platinum)	1600	1768	Yes
Pt–Rh (platinum–rhodium)	1700	1900	Yes
Mo (molybdenum)	2400	2623	No
Ta (tantalum)	2600	3017	No
W (tungsten)	2900	3422	No
C (graphite)	2600	3462	No
ZrO_2 (zirconia)	2000	2715	Yes

[a]Maximum working temperature in proper environment

Fe–Cr–Al heating elements are often called by the original trademark Kanthal, and are alloys of iron, chromium, and aluminum. There are many varieties from different manufacturers with slightly different alloy compositions. Kanthal can be heated to temperatures higher than nichrome, mainly due to the increased stability from the protective Al_2O_3 surface layer. Because the maximum working temperature of Kanthal heaters (about 1200 °C) is comparable to that of silica glass, they are useful for growth experiments using sealed silica tubes.

Both nichrome and Kanthal are resistive metallic alloys, and they have similar resistivities and other overall properties (although Kanthal is more brittle). Their lifetime is long if corrosion is prevented and temperatures are not permitted to reach or exceed maximum working temperature. Some commercial furnaces made of these heating elements are not designed to be repaired; it is best to avoid using these furnaces when volatile flux is involved.

Silicon carbide (SiC) heating elements are sintered ceramics available in the form of straight rods, U- and W-shaped rods, tubes, and spiral tubes (*see* Fig. 5.4 for some examples). SiC heating elements are useful for flux growth because (1) their maximum working temperature of about 1400 °C is sufficiently high to cover most experiments, (2) they can be used in both oxidizing and reducing environments, and (3) damaged SiC heating elements can be replaced easily at low cost. On the down side, the resistance of SiC heating elements increases gradually even when they are used properly. To offset this, the power controller of a SiC furnace is usually equipped with a switch to adjust the transformer voltage. The voltage is increased as the heating element becomes more resistive from aging and corrosion, and this process can be repeated until the resistance reaches the limit set by the power supply.

Molybdenum disilicide ($MoSi_2$) heating elements are often called Kanthal Super after its trademark; this can lead to some confusion, as Kanthal and Kanthal Super

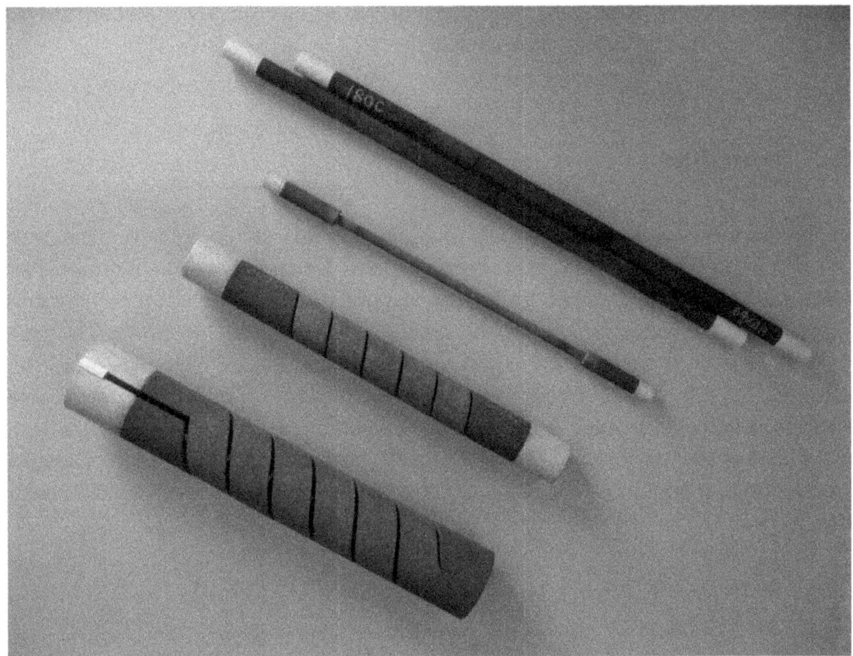

Fig. 5.4 SiC heating elements in various forms. Courtesy of Siliconit

are two completely different materials. $MoSi_2$ can be used to temperatures higher than SiC, but they are more expensive and require greater care in handling due to their brittleness. The stability of $MoSi_2$ at high temperatures comes from its protective SiO_2 layer, which forms during its first heating in air. Because the SiO_2 layer loses its structural integrity below 800 °C, $MoSi_2$ should only be used above this temperature. Under normal operating conditions, $MoSi_2$ has a long lifetime.

Lanthanum chromite ($LaCrO_3$) heating elements have a high maximum working temperature of 1800 °C in air. They give off Cr_2O_3 vapor at high temperatures, and are susceptible to thermal shock.

5.1.4 Summary of Furnaces

To summarize this section, any researcher involved in materials synthesis should benefit from owning at least one box furnace equipped with Kanthal heating elements, or failing that, a nichrome box furnace. Such a furnace is useful not only for crystal growth, but for many other operations requiring moderately high temperatures. For many flux growth experiments, SiC heating elements are an optimum choice, because they cover the appropriate temperature ranges and their cost of repair is relatively low. It must be kept in mind that a larger furnace is not necessarily a better furnace, because the demands on electric power and repair costs rise sharply with increasing furnace size.

5.2 Crucibles

In flux growth, the starting materials of solute and flux are completely melted into solution at high temperatures. This creates a need for crucibles, which must (1) not react with the solution or the growth atmosphere, (2) provide adequate structural support for the solution, and (3) be available at a reasonable price. The word "crucible" is used here for a container of any shape that is intended to hold the solution. To minimize corrosion, metallic crucibles are often used to grow ionic crystals, whereas crucibles made of ionic materials are frequently used to grow metallic crystals. Table 5.2 lists major crucible materials and their basic properties; among these, platinum, silica glass, alumina, and tantalum are most widely used in flux growth.

Table 5.2 Properties of typical crucible materials

Material	T_{max} (°C)[a]	T_{melt} (°C)	Use in air	Main use
Pt (platinum)	1500	1768	Yes	Oxides, halides
Ag (silver)	550	962	Yes	Hydroxides, halides
Au (gold)	950	1064	Yes	Halides
Ir (iridium)	2200	2446	No	Oxides, halides
Ni (nickel)	900	1455	No	Halides
Mo (molybdenum)	1900	2623	No	Metals, oxides, halides
Ta (tantalum)	2400	3017	No	Metals, halides
W (tungsten)	2500	3422	No	Metals, chalcogenides, oxides
C (graphite)	2000	3462	No	Metals, halides, chalcogenides
SiO_2 (silica glass)	1200	1700	Yes	Metals, halides, chalcogenides
Al_2O_3 (alumina)	1800	2050	Yes	Metals, hydroxides, halides
MgO (magnesia)	2200	2852	Yes	Metals
ZrO_2 (zirconia)	2300	2715	Yes	Metals, chalcogenides
BN (boron nitride)	1800	2973	No	Metals

[a]Maximum working temperature in proper environment

5.2.1 Platinum

Platinum crucibles are used for the flux growth of most oxides, simply because there is no alternative. Platinum is a member of the noble metals, and has a high melting point, high stability in oxidizing atmospheres (such as air), and resistance from many chemicals including acids and basic solutions. The melting point of platinum is 1768 °C, and the maximum working temperature is about 1500 °C. Platinum crucibles are also malleable enough to be bent by hand, which becomes a useful property during several stages of the experiment.

Unfortunately, platinum crucibles are expensive. This is mostly due to the high cost of the raw material. Unlike many other materials, the price of platinum changes on a daily basis and so does the price tag of platinum crucibles. (Some vendors sell platinum wares at fixed prices, but they usually charge a lot more than the market price.) As an example, the price of platinum hit an all-time high of US$72 per gram in March 2008, when the global financial crisis led investors to flood into platinum as a hedge against the crumbling stock market. The price then dropped to $25 in late 2008, before rising again to $60 in 2011. To convert these numbers into another currency, the exchange rate at that time must be considered.

Two aspects of platinum crucibles that are different from most other crucibles are that (1) they can be used repeatedly, perhaps ten times or more, if handled and cleaned properly, and (2) damaged crucibles can be refabricated at a fraction of the original price. Because of these two features, the running cost of using platinum crucibles is not much higher than those of other crucibles. The cleaning procedure for platinum crucibles will be described in Sect. 5.4.

The manufacturers of platinum crucibles are also involved in the business of buying and selling the raw material, as well as in the fabrication of various platinum parts. These manufacturers retrieve damaged crucibles, melt and purify them, and refabricate new crucibles. Because platinum is turned into liquid during the process, the source platinum can be of any shape, size, and purity. The price of such refabricated crucibles can be as low as 10% of the original purchase.

For flux growth, it is convenient to purchase platinum crucibles in one of the standard sizes. In Japan, they are 10, 15, 20, 25, 30, 40, and 50 ml (*see* Fig. 5.5 for some examples). Although larger crucibles can yield larger crystals, the effect is much weaker than linear—a change from 10 to 50 ml would probably only double the size (volume) of the crystals. The Japanese standard crucible with a lid weighs 1 g for each 1 ml; a 15 ml crucible weighs 15 g, and a 30 ml crucible weighs 30 g. The lid is designed to be crimped around the rim of the crucible. Such a seal is not airtight but is adequate for most purposes. Lids can also be prepared by cutting a platinum sheet with a thickness of about 0.1 mm.

New crucibles tend to weld with the lid on their first use; this can be prevented to some extent by annealing empty crucibles and lids at 1100–1200 °C for one hour, or by roughening the contact area with abrasive paper. It has been claimed that the annealing procedure also helps protect new crucibles against corrosion.

Fig. 5.5 Front, from left to right: platinum crucibles with volumes of 15 ml (with its lid), 15 ml (sealed), 30 ml (thick walled), and 100 ml (beaker shaped). Back, from left to right: 30-ml alumina crucible with cover, 50-ml alumina crucible, and large alumina container with cover. Far back: 50-ml glass beaker

Platinum reacts readily with many elements and non-oxide materials. For example, it becomes brittle upon contact with C, S, Se, Te, P, As, Sb, and Cl_2 at high temperatures, and it will alloy with Pb, Bi, Cu, and Sn. For this reason, platinum crucibles should not be used in a reducing environment, as many fluxes (especially PbO, PbF_2, and Bi_2O_3) will turn into metals and react with platinum (*see* Fig. 5.6). At high temperatures, any residue of organic matter can turn into carbon and attack platinum. Even silica (SiO_2) can react with platinum, so only stable refractory oxides such as alumina (Al_2O_3) or zirconia (ZrO_2) should be in contact with platinum at high temperatures. To hold hot platinum crucibles, only tongs that have tips covered in platinum should be used.

Platinum is sometimes alloyed with iridium or rhodium to modify its properties. Iridium has a higher melting temperature than platinum, such that the alloyed crucibles can be used up to higher temperatures. Alloying of platinum with rhodium improves the strength of crucibles. However, both iridium and rhodium can contaminate crystals, so alloyed platinum crucibles are rarely used in flux growth.

Fig. 5.6 Corroded platinum crucible

5.2.2 *Silica Glass*

Silica glass (fused silica) is made of pure SiO_2 and it is often used in the form of sealed tubes or ampoules to create a protective environment. It is not quite correct to call it "quartz glass," because quartz refers to one of the crystal polymorphs of SiO_2. Standard silica tubes of various diameters can be purchased from various suppliers. As long as the diameter is less than 20 mm and the thickness less than 2 mm, it is not difficult to seal silica tubes using an oxygen–hydrogen blow torch. (Smaller tubes can be sealed with an oxygen–propane torch.) Compared with soda-lime or borosilicate glasses such as Pyrex and Vycor, pure silica has much higher resistance to thermal shock. This allows silica glass to be heated and cooled quickly, so that even the total beginner can learn to seal silica tubes with a little practice.

The maximum working temperature of silica glass is about 1200 °C. Silica glass is stable against some low-melting-point metals, sulfides, selenides, and chlorides, but is attacked by many metals at high temperatures. If the flux solution reacts with silica, a crucible made of a more appropriate material is inserted into the silica tube.

The surface of new silica glass is often contaminated by dust and grease. These should be removed using a soft brush and liquid detergent, followed by being given thorough rinse in warm water. A dilute solution of nitric acid or sulfuric acid can be used to remove any remaining stains. Because salt in tap water can lead to devitrification (crystallization) of silica glass above 1000 °C, the final rinsing should be done with distilled water. For the same reason, silica glass should not be touched by bare hands.

Flame sealing of a silica tube is performed while the tube is connected to a vacuum line (Fig. 5.7). If the starting material becomes slightly volatile and attacks silica at high temperatures (as occurs for Al metal), about 0.2 atm of argon gas is introduced into the tube; it will become close to 1 atm at high temperatures and suppress the evaporation. During the sealing process, when the flame is softening the necked part of the tube, abrupt pulling of the tube must be avoided (the necked part will mostly collapse by itself under external air pressure). The sealed part of the tube should be rounded for maximum strength.

For those samples that react with silica but not with carbon, the inner surface can be coated with a layer of carbon. This is done by washing the inside of a half-closed tube with an organic liquid, such as acetone or ethanol, and heating rapidly with a flame at about 700 °C. Because there is insufficient oxygen inside, the liquid cannot burn off and forms a pyrolyzed carbon layer on the wall. This process can be repeated several times to give a thick coating. To avoid contamination, only pure liquid should be used for this purpose.

5.2.3 Alumina

Alumina (polycrystalline Al_2O_3) crucibles are often used for the flux growth of various intermetallic compounds, as well as for oxides when a chloride or hydroxide is used as the flux. Alumina is not attacked by many of the low-melting-point metals used as a flux, such as Al, Cu, Zn, Ga, Ge, In, Sn, Sb, Pb, and Bi [2]. Although a pure melt of rare-earth metals reacts with alumina, about ten atomic percent of rare earth can be added to the low-melting-point metals and held in an alumina crucible, without attack, up to about 1200 °C [2].

Standard alumina crucibles are available in various shapes and sizes, which can be purchased with or without top covers. Only those with the highest purity (>99%) should be used for flux growth. Very small crucibles that will fit inside a silica tube can be difficult to find, and it may be necessary to order custom-sized crucibles for this purpose. It is possible to cut alumina crucibles with a diamond wheel saw, but with much more effort compared with cutting silica glass.

Flame sealing of alumina is extremely difficult, especially when plasma flames are not available. However, if the seal does not have to be completely airtight, pieces of alumina can be glued together using alumina cement; it is applied like regular glue, and a heat gun or powerful hair dryer is used to dry the cement for pre-hardening. The final hardening occurs when it is heated in air inside a furnace. Alumina cement is useful in sealing alumina containers that are used for growth in air.

Because molten chemicals clog grain boundaries, there is no satisfactory method of cleaning the used alumina crucibles in most cases. If an acid is used to clean such crucibles, it would cause cracks during the next heating. This is a common problem among ceramic materials, such as MgO, ZrO_2, and BN.

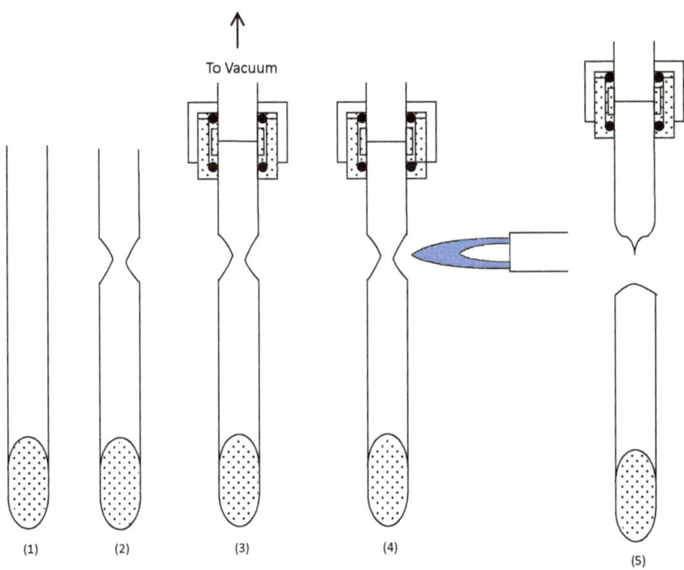

Fig. 5.7 Flame sealing of silica glass tube. When sealing volatile materials, the bottom of the tube is kept cold by immersing it in water or liquid nitrogen

5.2.4 *Tantalum*

Tantalum crucibles are sometimes used for the flux growth of intermetallic compounds, when the metallic solution attacks alumina crucibles. Tantalum crucibles can be sealed by arc welding, but since tantalum itself cannot be heated in air, it must be sealed in a silica tube or heated in a vacuum, neutral, or reducing environment. Tantalum crucibles can also be used to grow some oxides with reduced valence states under reducing conditions.

A tantalum crucible can be prepared by welding a cap onto the end of an open tube, or drilling a hole into a cylinder. Tantalum is used not because it is the most stable material against corrosion, but because it is very machinable; tungsten has a higher melting point and often shows less reactivity, but its extreme hardness makes it a formidable material to work with.

5.3 Starting Materials

In this section, we provide some of the basic information on the chemicals used as starting materials. Because it is not practical to list safety protocols and potential health hazards for every chemical used in flux growth, such details must be sought in appropriate places (such as a manufacturer's website).

In most cases, high-purity chemicals are purchased, rather than synthesized by the crystal grower. However, the purity shown on the label can be deceptive, as the number refers mostly to metallic impurities only. Even a 99.9% pure metal can contain several atomic percent of oxygen, where oxygen is either dissolved in the metal or present as oxide particles in the metal matrix. Similarly, unknown amounts of water can be included in hygroscopic oxides and halide salts, something which is not evident from the purity value shown on the label.

5.3.1 Chemicals Used in Oxide Growth

Most chemicals of oxides and halide salts are in the form of powders or small granules. If the chemicals are available in various particle sizes, it is reasonable to choose the largest one; very fine powders have large surface areas and can readily absorb water from the atmosphere. Regardless of the particle size, chemicals should be kept in tightly sealed bottles and stored in a cabinet, or in vacuum desiccators for better protection against moisture. It is a good practice to perform X-ray powder diffraction and thermogravimetry on any chemical of questionable quality, as these tests provide useful checks on the purity and composition of the material. Oxygen off-stoichiometry is particularly prevalent among the first-row transition metal oxides, and any departure from the expected value can cause errors in weighing out chemicals.

Chemicals in the most stable states are usually used for growth in air, but this point may not be explicitly stated in the literature. For example, when an oxide of alkali metal or alkali-earth metal appears as a component of flux growth, it is generally implied that the corresponding carbonate, such as Li_2CO_3 or $CaCO_3$, is used as the starting material. This is because the carbonates are more stable than oxides at room temperature (oxides quickly absorb moisture and CO_2), and carbonates decompose to become either oxides or bare cations when heated above 800–900 °C. Similarly, it is not necessary to use a metal oxide having the same valence as the one in the final crystals, and the most stable form (such as Fe_2O_3 and Tb_4O_7) is usually used as the starting material.

To use hygroscopic or CO_2-absorbing chemicals, it is often necessary to dry the powder before weighing. For stable oxides, drying can be done in air by heating in a box furnace for several hours (for example, at 900 °C for La_2O_3). In most cases, alumina crucibles can be used for drying chemicals below 1000 °C. For halide salts, it is often necessary to dry the powder using a vacuum oven.

5.3.2 Metals

The metallic elements used for the growth of intermetallic compounds come in a variety of forms (*see* Fig. 5.8 for some examples). It is sometimes necessary to cut

Fig. 5.8 Metallic elements in various forms

or break the metals into smaller pieces. For soft metals such as In, Pb, and Sn, large chunks can be cut with a clean knife or wire cutter. Brittle materials such as Te, Ge, and Sb can be broken into smaller pieces with a mortar and pestle. For harder metals such as the transition metals, they must be sawed and then etched before weighing.

If the surface of the metal is oxidized, etching is performed to remove the surface layer. Suitable etchants for each metal are described in metallurgy handbooks, but in most cases, they are an acid or a combination of acids. For example, nitric acid diluted by the same amount of water can be used for the rare earths and many other metals. The sample must be rinsed thoroughly after etching, and acetone is often used to dry off the surface at the end. Obviously, etching is easier if the sample is in the form of larger pieces.

To a large extent, the method of handling a metallic element depends on its reactivity with the atmosphere. The alkali metals are generally the most reactive, followed by the alkali earths, the rare earths, the transition metals, and other main-group metals.

The alkali metals (Li, Na, K, Cs, and Rb) react strongly with air and water. They are often stored in oil, but storage in argon-filled ampoules is preferable. To perform

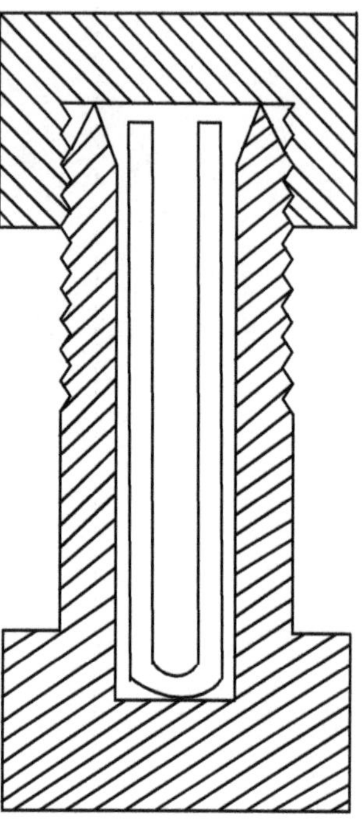

Fig. 5.9 Stainless steel container with a screw-fitting thread, which can be used to seal an alumina crucible by hand in a glove box

all preparation processes inside a glove box, iron or stainless-steel containers with a screw-fitting thread (Fig. 5.9) have been devised [3, 4]. Such a container can be sealed airtight by hand, pre-empting the need to take the sample out of the glove box for the sealing procedure. If an arc welder is connected to the glove box, the sealing of tantalum tubes can be carried out without exposing the sample to the atmosphere.

The alkali-earth metals (Ca, Sr, and Ba) are sometimes stored in oil. The oil is removed with an organic solvent such as hexane, and large metal pieces can be cut into smaller sizes with a clean wire cutter. As these metals react with water, they cannot be etched with acids. They are handled in a glove box, but may be exposed to dry air for a few seconds. The rare-earth metal Eu has similar properties to alkali-earth metals and is handled in a similar manner.

The reactivity of rare-earth metals with air varies from the most reactive Eu to the least reactive Sc. The exact order is Eu, La, Ce, Pr, Nd, Sm, Yb, Gd, Tb, Y, Dy, Ho, Er, Tm, Yb, Lu, and Sc, and except for Yb, this is the order of decreasing metallic radius [5]. Even the reactive members, La, Ce, Pr, and Nd, can be handled briefly in air, but these metals will be oxidized completely if left in air. The recommended method of storage is to seal the metal in an evacuated glass ampoule.

Most transition metals can be handled in air, but powders and small pieces should be stored in sealed ampoules for prolonged storage. The same is true for relatively stable main-group metals.

5.4 Growth of Oxide Crystals in Air

Having looked at the equipment and chemicals in some detail, let us now examine flux growth experiments in action. In this section, we describe a typical experiment in which a platinum crucible is used for flux growth in air. Thus, the techniques apply mostly to oxide growth. No mention is made of advanced techniques such as seeding and crucible rotation, as these are rarely used for the growth of the millimeter-sized crystals needed for physical measurements. Flux growth in protective atmospheres will be discussed in Sect. 5.5; however, some of the basic procedures common to both types of experiments are not repeated there. It is assumed that a fume hood is available for the handling of acids and basic solutions, and furnaces are located in a well-ventilated area where fumes can escape without coming into contact with workers in the room. Figure 5.10 shows some of the tools used in flux growth experiments.

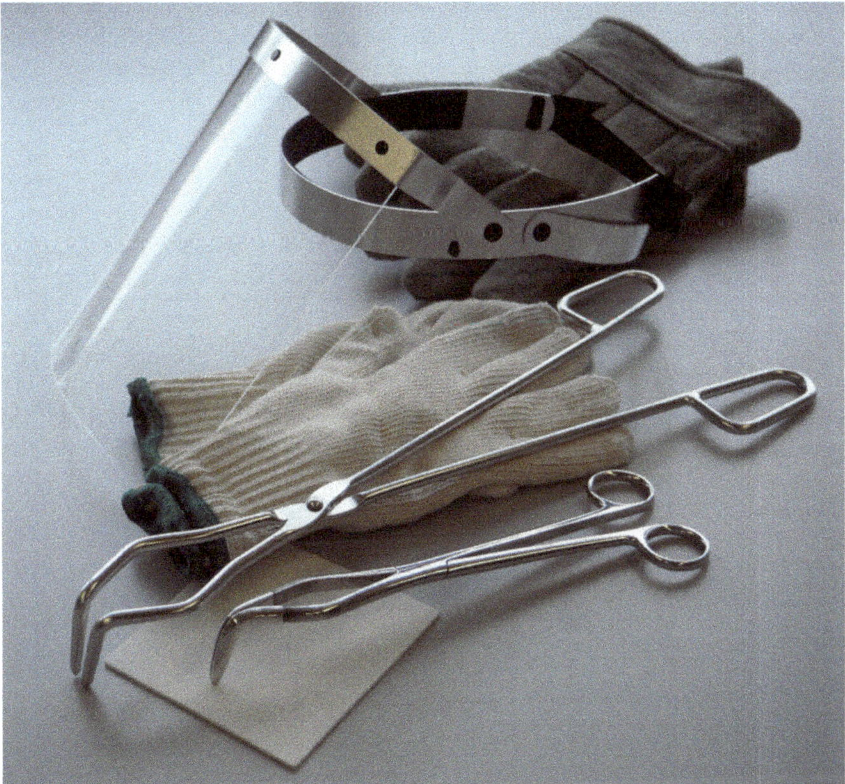

Fig. 5.10 Some of the tools used in flux growth. The smaller tongs have platinum-covered tips for holding hot platinum crucibles. The gloves at the front are made of 100% cotton and do not melt from heat like polyester gloves. For handling at high temperatures, the gloves at the back are used

5.4.1 Preparation

Preparation begins with weighing the starting chemicals in powder form, followed by thorough mixing in a clean bottle. (In some cases, it is helpful to presynthesize the desired compound in a polycrystalline form, for which the mixing is carried out with a mortar and pestle.) It is difficult to predict the amount of chemicals that will fill a given size of crucible, but as a rough estimate, powder that fills 70% of a 50-ml bottle should be a suitable amount for use in a 15-ml crucible.

The mixed powder is then loaded into the crucible. Even when the powder is pressed into the crucible, there will be large reduction in volume once the powder is molten. To fill the crucible as much as possible, premelting of the initial charge followed by refilling with more powder is performed. The premelting should be carried out in a box furnace, where the temperature is often set to 850–950 °C. Repeating premelting two or three times usually fills the crucible to about 80% of its height.

To avoid spilling the premelt, the platinum crucible should be covered and placed in an alumina container. Gentle heating can prevent the abrupt bursting of gas bubbles, which occurs especially when carbonates are included in the mixture. If liquid escapes out of the crucible by creeping up the wall, the premelt procedure

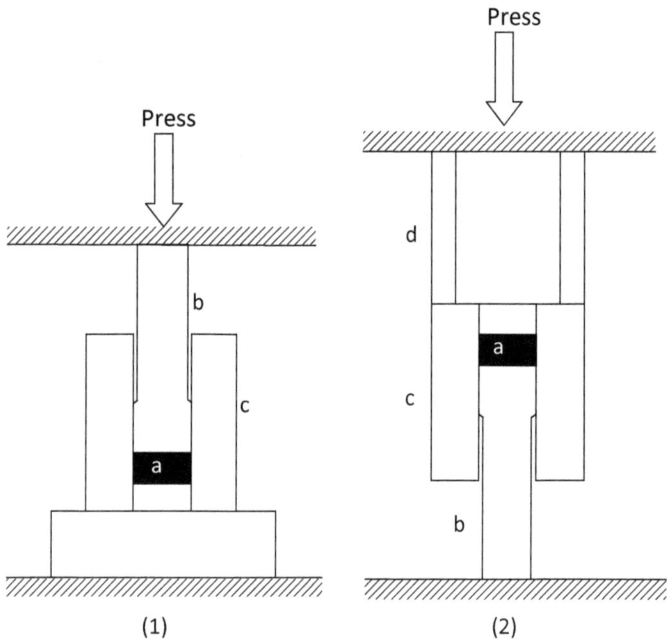

Fig. 5.11 Preparation of compacted tablets using a hydraulic hand press and hardened steel die. a: sample; b: piston; c: cylinder body; d: ring. The ring is used to remove the tablet from the cylinder body

should be changed to filling the crucible only once with highly compressed powder. If a hydraulic hand press is used to prepare hard tablets (*see* Fig. 5.11), up to about 70% filling can be achieved without premelting.

After a sufficient amount of sample has been charged into the crucible, it is usually sealed with a platinum lid. If a volatile flux is used, simply resting the lid on the crucible does not provide an adequate seal; instead, the lid must be crimped into place around the entire rim. The standard lid that comes with the crucible fits perfectly, but a similar lid can be made by cutting a platinum sheet. For a perfect airtight seal, the lid must be welded onto the crucible using a blow torch or arc welder.

The sealed crucible is then placed inside a larger alumina crucible. The space between the platinum crucible and the alumina crucible should be filled with alumina powder, and then an alumina lid is sealed onto the alumina crucible using alumina cement. These protective measures help to prevent flux vapor from escaping and contaminating the furnace. Also, the added thermal mass damps temperature fluctuations during the slow cooling for crystal growth.

The crucible is then loaded into the furnace. If a vertical tube furnace is used for the growth, silica wool should be tightly packed into the top part of the tube. This will trap most of the flux vapor that escapes from the crucible, with the vapor condensing back into powder.

5.4.2 Growth

The temperature profile of the entire growth sequence is programmed using the temperature controller. Typical growth involves a heating rate of 200–300 °C per hour to the highest temperature, soaking at the highest temperature for a period of 2–24 h, and slow cooling to a specified temperature at a rate of between 0.5 and 5 °C per hour. The final temperature should be maintained for a certain period if the experiment requires decanting of the liquid at the end. Otherwise, the heater power is turned off at this stage and the crucible is cooled to room temperature.

The soaking period must be long enough for complete dissolution to take place, as undissolved solute will act as nucleation centers during crystallization. The soaking temperature should be at least 50 °C above the liquidus temperature. However, setting the maximum temperature too high can have adverse effects, such as increased volatility and increased reactivity toward the crucible. Also, undesired phases may be stabilized if the temperature is set too high.

During slow cooling, nucleation starts after the temperature drops below the liquidus point. There are several techniques that can help to reduce the number of nuclei, and thereby increase the size of the final crystals. One technique involves the use of temperature oscillation at the beginning of slow cooling. For example, 10 °C oscillations with a 10 min period are added to an average cooling of 1 °C per hour. The idea here is to redissolve small crystals that nucleate during the initial cooling, leaving fewer larger crystals for further growth. Another method of increasing crystal size is to localize the nucleation site. With a vertical tube furnace, this can be done by blowing air onto the bottom of the crucible. Sending in fresh air also helps to maintain an oxidizing environment, which can protect the platinum crucible from corrosion.

To grow large and high-quality crystals, the rate of cooling must be slow enough to avoid exceeding the maximum rate of stable growth. In principle, the cooling rate should be increased in a nonlinear manner, corresponding to the rate at which the surface area of crystals increases with time [6]. However, this is not possible with most temperature controllers, and a reasonable compromise is to set a series of linear cooling rates. (For example, 1 °C per hour from 1250 to 1150 °C, 1.5 °C per hour from 1150 to 1050 °C, and 2.0 °C per hour from 1050 to 950 °C.) In practice, the cooling rate is often simply kept at a single value for the entire duration.

5.4.3 Removal of Crystals

Slow cooling is stopped before the temperature reaches the eutectic point. If a solvent that dissolves or leaches the solidified flux, but not the crystals, is known, then the crucible can be removed from the furnace at room temperature. A dilute solution of nitric acid is probably used most often for leaching the flux; to speed up the process, the dissolution is usually carried out on a hot plate. In many cases, the solidified flux becomes brittle after some leaching, and the crystals can then be picked up with a pair of tweezers. If the crystals are firmly attached to the crucible wall, they can be separated by bending the wall away from the site of attachment. This must be done carefully, as damage to the crystals can be caused by impatience.

Sometimes the solidification of flux leads to the straining and cracking of grown crystals. There are also instances where no solvent that dissolves only the flux component can be found. In these cases, the residual liquid must be decanted at the end of slow cooling [7]. This can be done by using tongs to remove the crucible from the furnace, pouring off the excess liquid, and placing the crucible back into the furnace (to avoid thermal shock to the crystals). Obviously, this method is not possible if the lid is tightly sealed to the crucible. For sealed crucibles, decanting using a centrifuge can be useful, and this technique is described in the next section on the growth of intermetallic compounds. If the crystals cannot be separated from the flux through leaching or decanting, it may be necessary to chisel away the solidified flux using a small drill.

5.4.4 Cleaning of Platinum Crucibles

If the platinum crucible did not suffer significant damage, it can be cleaned for another use. The crucible can be cleaned by melting a convenient and non-toxic flux, such as Na_2CO_3, inside the crucible. After a good soak at a high temperature

(950 °C for Na_2CO_3) in a box furnace, the liquid is decanted into a used alumina crucible outside the furnace. This process is repeated until no impurity is visible in the decanted flux. The crucible is then immersed in a solution of dilute nitric acid, where, when basic Na_2CO_3 is used, the neutralization reaction helps to remove any residue on the crucible wall. Any remaining stain on the crucible can be removed by adding hydrochloric acid, at 1–2%, to the nitric acid solution. The crucible should be rinsed thoroughly with distilled water before drying.

If a small hole is formed on a platinum crucible, it can be closed using the following procedure. First, the crucible is rested firmly on a ceramic plate, and a piece of platinum is placed to cover the hole—such a piece can be obtained by cutting the edge of a platinum lid. Next, the platinum piece is heated with a blow torch. As the piece starts to melt, it first becomes rounded and then turns into a sphere. Then, the surrounding area also starts to melt, and in an instant, the sphere spreads over the hole, becoming part of the crucible. It is important to stop the heating at the right moment, as any further heating will create a hole larger than the original.

If the result of such a repair is less than satisfactory, the crucible should be sent for refabrication since there is too high a risk of leakage if damaged crucibles are used.

5.5 Flux Growth in a Protective Atmosphere

Roughly speaking, dry air is composed of 78% nitrogen and 21% oxygen, with argon and carbon dioxide taking up much of the remaining 1%. As nitrogen gas is relatively inert, air is basically an oxidizing environment, providing an appropriate growth atmosphere for many oxides. On the other hand, for compounds that are not stable in air at growth temperatures, some form of controlled environment must be provided.

One way to control the growth atmosphere is to introduce a flow of an appropriate gas, with sealing fixtures attached at the open ends of a tube furnace (*see* Sect. 5.1). With flowing argon gas, for example, the inert atmosphere needed for the growth of intermetallic and other non-oxide compounds can be obtained. Similarly, by controlling the oxygen partial pressure, oxides having metal ions in reduced valence states can be grown in a tube furnace.

Except for the handling of gases, the basic procedures involved in these experiments are similar to those mentioned in the previous section. Therefore, in this section, we describe another method of providing a protective atmosphere, which is to seal the crucible in a silica glass tube (*see* Sect. 5.2 for the properties of silica glass). Compared with flowing gases, using a sealed silica tube has the following advantages:

1. Simple box furnaces can be used for the experiment, and the growth of many batches can be carried out in a single run.
2. Contamination from impurities in flowing gases and from volatile components of ceramic tubes can be avoided.

3. Grown crystals can be separated from the residual flux by decanting in a centrifuge. This is especially convenient when the flux cannot be leached away from the crystals, as is frequently the case in the growth of intermetallic compounds.

The disadvantages are that only small crucibles that fit inside the silica tube can be used, and the maximum growth temperature is limited to about 1200 °C.

Historically, the flux growth of intermetallic compounds gained much popularity after the technique of decanting in a centrifuge was reviewed by Fisk and Remeika [1]. Various refinements and modifications to this technique have been reported in a number of subsequent papers [2, 8–11]. This section can be regarded as a brief summary of these papers.

5.5.1 Use of a Crucible Inside a Silica Glass Tube

If the starting chemicals do not react with silica, the sealed silica tube itself can be used as the crucible. For example, the growth of selenides and tellurides from a chloride flux is often carried out directly in silica tubes. On the other hand, when elements such as Al, Si, the alkali earths, the rare earths, and transition metals are involved, it is usually necessary to place a crucible of a different material inside the silica tube. Possible crucible materials include alumina, tantalum, and boron nitride, with alumina being used most often for the flux growth of intermetallic compounds.

The growth experiment starts with the preparation of a silica tube with one of its ends closed. This is done by heating the middle of a long tube using a blow torch that is fixed to the glass bench. To prevent the tube from cracking during the growth, the closed end should be rounded or flat-bottomed, rather than tapering and sharply pointed.

The starting chemicals are then weighed and placed inside an alumina crucible. It is important to handle the chemicals in such a way as to minimize contamination; for those chemicals that are sensitive to air, the process should be carried out in a glove box. Because metals are usually in the form of pellets rather than fine powders, the starting chemicals are not thoroughly mixed. Instead, chemicals with low melting temperatures are placed on top of high-melting-point chemicals in the crucible. This way, the low-melting-point chemicals will flow over and swallow other chemicals as melting progresses.

The filled alumina crucible is then inserted into the silica tube, on top of silica wool at the bottom (*see* Fig. 5.12). Additional silica wool is plugged into an empty alumina crucible, which is then turned upside down and inserted into the silica tube. This is the catch crucible, which acts as a sieve to capture grown crystals in the decanting process. If the decanted liquid does not severely attack the silica tube, silica wool can be inserted into the silica tube without using an empty alumina crucible. Alternatively, replacing the silica wool with a fritted alumina disk placed between the two crucibles can enable cleaner decanting in some cases [11]. To

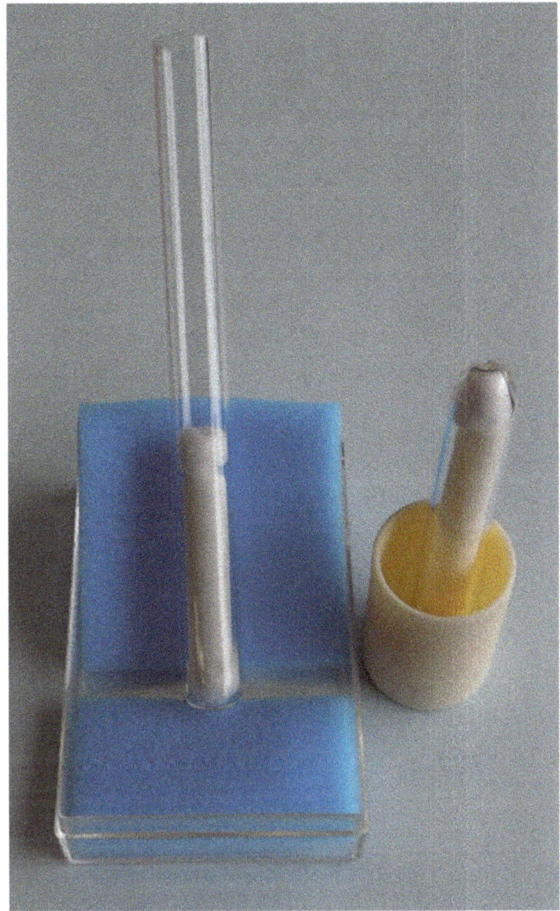

Fig. 5.12 Growth and catch alumina crucibles in a silica glass tube. Left: unsealed tube. Right: vacuum-sealed tube, placed in an alumina container

prevent the crucible from smashing the silica glass tube, silica wool should be placed at the very top of the tube. The silica tube is then flame-sealed in a vacuum line, by the method described in Sect. 5.2.

For chemicals that react with alumina, or for those with high vapor pressures that will attack silica, sealed tantalum crucibles are often used as an alternative. The following process can be used to prepare a tantalum crucible [8, 12]. First, a tantalum tube and three caps (made from a tantalum sheet) are prepared. Small holes are drilled into one of the caps, which will act as a sieve. One of the remaining two caps is then arc-welded onto the tube to create a crucible. After the chemicals are placed into the bottom of the crucible, the drilled cap is placed inside, just above the level of the chemicals. Finally, the crucible is sealed by welding the third cap onto the open end. Because tantalum reacts with air at high temperatures, the tantalum crucible must be sealed in a silica glass tube.

The sealed silica tube is kept nearly upright by placing it in an alumina container (Fig. 5.12) that is then put into a box furnace. The cooling rate used for the growth of intermetallic compounds is usually about 3–10 °C per hour; the lower viscosity of metallic melt allows a higher cooling rate than for oxide growths. When slow cooling is completed, a pair of tongs is used to remove the alumina container from the furnace, and the container is inverted to slide the silica tube into one of the metal cups of a centrifuge. The centrifuge is then turned on for several seconds for the decanting to take place. The cooled silica tube is then wrapped in a thick paper towel and opened by tapping with a hammer.

To prevent the silica tube from shattering in the centrifuge, a thick layer of silica wool should cover the bottom and sides of the metal cup. Placing a counterweight in the cup on the opposite side (there are usually two or four cups in a centrifuge) helps to maintain a smooth spin. Because the decanting takes place in the first few seconds, a prolonged time of spinning only increases the risk of breaking the silica tube.

In this section, we covered only the basic procedures of growth in a sealed silica tube. The references cited above describe various techniques that have been used in the flux growth of intermetallic compounds.

5.6 Some Notes on Handling the Grown Crystals

To be used as samples for physical measurements, grown crystals must be properly assessed and characterized. In most cases, characterization involves studying the structure and composition, as well as the impurities present in the crystals. X-ray powder diffraction is used most often to check phase purity and crystal structure. Instruments such as electron probe micro-analyzers (EPMAs) provide detailed mapping of the elements present in both the main crystal phase and possible flux inclusions. (For solid solutions in particular, a detailed compositional analysis is essential.) Rather than attempting to cover the vast subject of characterization, this section offers some tips on basic procedures that do not require any special equipment.

For the crystal grower, characterization starts as soon as crystals are found in the crucible. Photographs should be taken while the crystals are still in the crucible, as they often turn out to be a useful record. The first observation should also be written down in a way that later becomes the standard. This way, the results of different experiments can be compared, and it avoids any gaps in information normally noted but briefly forgotten in a particular instance. Important data include the size, number, habit, and location of the grown crystals, as well as the amount of solidified flux remaining in the crucible. If crystals of different compounds are found alongside the main crystals, these should also be recorded.

Once the crystals have been extracted and cleaned, it is time to look at them under a microscope. As discussed in Chap. 2, much information can be obtained by simply looking at the crystal in the as-grown state. A straight line found on a natural face can be a sign of a twin plane. Growth hillocks and hollow cores are usually evidence of dislocations or stacking faults. If the crystal surface is treated with a suitable etchant, etching occurs preferentially at dislocation sites. The cavities formed by etching are called etch pits, and their shape often reflects the underlying crystal structure. The dislocation density can be determined by counting the number of etch pits per unit area.

The interior can also be inspected if the crystal is transparent to visible light. However, trouble is often experienced because the illuminating light is reflected off the crystal surface. This can be greatly reduced by placing the crystal in a small Petri dish filled with immersion oil or water. (Surface reflection is completely eliminated if the crystal is immersed in a liquid having the same refractive index.) If the crystal is illuminated from the sides or below with low-intensity light, flux inclusions can be seen clearly as dark spots on a light background. Additional defects and strains can be detected if polarized light is used. Light guides made of fiber optics are useful in lighting from various directions.

Flux-grown crystals are usually handled with tweezers. For someone who is not used to working with tweezers, some practice can prevent damaging or losing precious crystals. To hold a small crystal, it is first necessary to pick up the crystal in the correct position so that it is secure in the tweezers. Then just the right amount of pressure is applied to the arms to retain the crystal. If too much pressure is applied, a poorly positioned crystal can fly out of the tweezers. Metallic tweezers with soft plastic tips are especially useful in handling brittle crystals. As cheap tweezers often have bad alignment and make the handling process extremely stressful, it makes sense to invest in a decent pair of tweezers.

When crystals are not in use, they should be kept in a labeled container. Keeping things tidy and in order around the work bench is also important. Many workers know how easy it is to pick up a crystal and then think about something else, forgetting where the crystal was originally placed and what other crystals were with it. Sorting out mixed-up crystals can be more time-consuming than the crystal growth itself. Finally, a detailed record of all the crystals should be kept on a permanent basis. Although it may seem superfluous at the time, the crystals can be of considerable interest to others in the future. As an extreme example, in 1969 Nassau was able to study various rubies grown in the 19th century, only because the crystals had been recorded and preserved in museums or private collections [13, 14].

References

1. Z. Fisk, J.P. Remeika, in *Handbook on the Physics and Chemistry of Rare Earths*, vol. 12, ed. by K.A. Gschneider, Jr., L. Eyring (Elsevier, Amsterdam, 1989), pp. 53–70
2. P.C. Canfield, in *Properties and Applications of Complex Intermetallics*, ed. by E. Belin-Ferré (World Scientific, Singapore, 2010), pp. 93–111
3. J. Akimoto, H. Takei, J. Solid State Chem. **85**, 31–37 (1990)
4. K. Kihou, T. Saito, S. Ishida, M. Nakajima, Y. Tomioka, H. Fukazawa, Y. Kohori, T. Ito, S. Uchida, A. Iyo, C.-H. Lee, H. Eisaki, J. Phys. Soc. Jpn. **79**, 124713 (2010)
5. B.J. Beaudry, K.A. Gschneidner, Jr, in *Handbook on the Physics and Chemistry of Rare Earths*, vol. 1, ed. by K.A. Gschneidner Jr., L. Eyring (Elsevier, Amsterdam, 1978), pp. 173–232

6. D. Elwell, H.J. Scheel, *Crystal Growth from High-Temperature Solutions* (Academic Press, London, 1975)
7. T. Wolf, Philos. Mag. **92**, 2458–2465 (2012)
8. P.C. Canfield, Z. Fisk, Philos. Mag. B **65**, 1117–1123 (1992)
9. P.C. Canfield, I.R. Fisher, J. Cryst. Growth **225**, 155–161 (2001)
10. C. Petrovic, P.C. Canfield, J.Y. Mellen, Philos. Mag. **92**, 2448–2457 (2012)
11. P.C. Canfield, T. Kong, U.S. Kaluarachchi, N.H. Jo, Philos. Mag. **96**, 84–92 (2016)
12. A. Jesche, P.C. Canfield, Philos. Mag. **94**, 2372–2402 (2014)
13. K. Nassau, J. Cryst. Growth **5**, 338–344 (1969)
14. K. Nassau, *Gems Made by Man* (Chilton Book, Radnor, Pennsylvania, 1980)

Chapter 6
Examples of Flux-Grown Crystals

I hope this book has served as a good starting point, an introduction, to the principles and techniques of flux crystal growth. Up to this point, only the growth of crystals has been emphasized—their interesting physical properties were rarely mentioned. To make this book more complete, and perhaps even inspiring, this final chapter presents 12 examples of flux-grown crystals. These examples are chosen mainly because the growth conditions for high-quality crystals are well documented in the literature; in each case, the basic recipe used to grow the crystals is described in the figure caption. (The size of the platinum crucible was either 15 or 30 ml.) The comments in each section are self-contained and the following 12 sections can be read in any order. Many terms used in this chapter are not explained, and the reader is encouraged to delve into the literature to learn more about the topic. In each figure, the smallest division is 1 mm.

6.1 $BaFe_2As_2$

Single crystals of $BaFe_2As_2$ are shown in Fig. 6.1.

An important aspect of high-temperature copper oxide superconductors is the presence of many chemical and structural variants. Basically, as long as the structure of the planar copper–oxygen square lattice is maintained, researchers can modify the composition and structure to a large extent, and study how the superconductivity is affected. This flexibility is one of the reasons why so many researchers studied this topic after its discovery in 1986.

Although other superconductors with interesting properties were discovered since copper oxides, many of them tended to be "one of a kind" or nearly so. For example, Sr_2RuO_4 is an extremely interesting superconductor [2]. However, among other reasons, the lack of related superconductors limited the number of researchers studying this compound. What can happen in such a case is that only a few devoted physicists study the compound into ever deeper levels, leaving behind other scientists.

M. Tachibana, *Beginner's Guide to Flux Crystal Growth*, NIMS Monographs,
DOI 10.1007/978-4-431-56587-1_6

Fig. 6.1 $BaFe_2As_2$. Ba and FeAs in 1:4 molar ratio were cooled from 1180 to 1000 °C over a 36-hour period in an alumina crucible sealed inside an argon-filled silica glass ampoule. The excess liquid was decanted using a centrifuge. Based on [1]

This is why the 2008 discovery of high-temperature superconductivity in iron-based compounds [3] led to great excitement. Not only was the transition temperature raised to 56 K, but also nearly 100 different superconductors were quickly identified. The nature of the superconductivity is unconventional [4], which makes these compounds all the more interesting. A major drawback, however, is the presence of toxic arsenic in many of the compounds.

$BaFe_2As_2$ is a prototypical member of iron-based superconductors. Although $BaFe_2As_2$ itself is not a superconductor at normal pressure, superconductivity can be induced by replacing some of the Ba, Fe, or As ions with a suitable element, or by applying a strong stress on flat faces of the crystals.

6.2 $CdCr_2Se_4$

Single crystals of $CdCr_2Se_4$ are shown in Fig. 6.2.

Only a few compounds are ferromagnetic semiconductors; most ferromagnets are metals, and most magnetic semiconductors undergo antiferromagnetic order. Nevertheless, ferromagnetic semiconductors play important roles in spintronics, a technology that combines electronics with the manipulation of mobile electron spins.

Fig. 6.2 $CdCr_2Se_4$. Cd, Se, and $CrCl_3$ in 4:4:2 molar ratio were cooled from 900 to 500 °C over a seven-day period in an evacuated silica glass ampoule. Based on [5]

Spinel $CdCr_2Se_4$ is a ferromagnetic semiconductor with a Curie temperature of 130 K. The ferromagnetism comes from the ferromagnetic Cr^{3+}–Se–Cr^{3+} superexchange interaction, which is slightly stronger than the antiferromagnetic Cr^{3+}–Cr^{3+} direct exchange interaction; this is a subtle issue, as antiferromagnetic order is found in isostructural $CdCr_2O_4$. Although $CdCr_2Se_4$ is not currently studied actively for device applications, the large magneto-electric effects found near the Curie temperature provide interesting subject matter for transport and optical studies [6].

Because $CdCr_2Se_4$ decomposes into CdSe and Cr_2Se_3 before reaching its melting point, crystals cannot be grown from melt. In addition to the flux method, single crystals of $CdCr_2Se_4$ can be grown using chemical vapor transport [7]. Typical transport agents are $CrCl_3$ and Cl_2; it is easier and safer to use $CrCl_3$, which is solid at room temperature.

6.3 $CuGeO_3$

Single crystals of $CuGeO_3$ are shown in Fig. 6.3.

In the study of magnetism, exotic behavior is often observed in low-dimensional systems with a low spin quantum number, such as spin-1/2. Because of the enhanced quantum fluctuations in these systems, conventional long-range magnetic order is replaced by a variety of novel quantum behaviors [9]. Low dimensionality also makes theoretical and numerical studies easier than in the case of three dimensions, allowing close comparison between theory and experiment.

The spin-Peierls transition is one such example, and occurs in antiferromagnetic Heisenberg spin-1/2 spin chains that are strongly coupled to three-dimensional lattice vibrations. Below this transition, the structural lattice is deformed and two adjacent spins combine to form a non-magnetic entity called a spin singlet or dimer. The first inorganic material found to show a spin-Peierls transition is $CuGeO_3$ [10]. Although several organic magnets were known to undergo spin-Peierls transition,

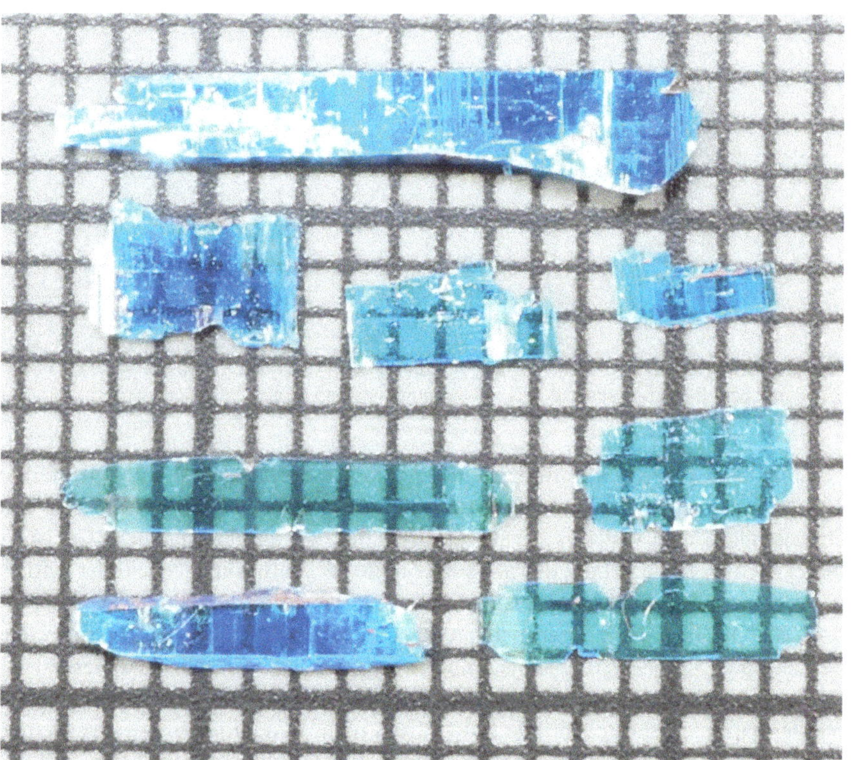

Fig. 6.3 $CuGeO_3$. CuO and GeO_2 in 0.15:0.85 molar ratio were cooled from 1160 to 1100 °C over a 70-hour period in a covered alumina crucible. $CuGeO_3$ can also be grown from excess CuO. Based on the phase diagram in [8]

its discovery in $CuGeO_3$ provided an opportunity for a complete experimental characterization of this transition. This is because the effects of impurities can be easily studied with $CuGeO_3$, and large crystals are readily obtained for this simple inorganic compound. Low-dimensional quantum magnetism remains an active field of study and is often invigorated by the discovery of new model materials.

6.4 $Dy_2Ti_2O_7$

Single crystals of $Dy_2Ti_2O_7$ are shown in Fig. 6.4.

At first thought, ice does not seem like an exciting material—even children know that ice is made up of H_2O molecules. Yet, a closer look at its crystal structure reveals the remarkable physics of this solid. In ice, each oxygen ion is surrounded by four other oxygen ions. These four oxygen ions form a tetrahedron, with the initial oxygen at the center. There is one hydrogen ion (proton) between the center oxygen and corner oxygen, so there are four protons in the tetrahedron. In order to form a stable H_2O molecule, two of the four protons must be closer to the center

Fig. 6.4 $Dy_2Ti_2O_7$. Dy_2O_3, TiO_2, MoO_3, PbF_2, PbO, and PbO_2 in 1.8:0.8:5.0:7.0:0.5 weight ratio were cooled from 1250 to 800 °C at a rate of 1 °C per hour in a hand-sealed platinum crucible. Based on [11]

oxygen, while the other two must be farther from the center. Any two of the four protons can be the closer ones. In a crystal of ice, the tetrahedron is repeated many times: the corner oxygen becomes the center oxygen in the adjacent tetrahedron, and the same rule of two "near" and two "far" protons is again enforced. Pauling in 1935 showed that there is no unique way for the protons to order in ice, leading to a vast number of energetically equivalent proton arrangements [12]. As a result, there is frozen-in disorder of protons in ice.

The same exact physics is observed in the cubic pyrochlore compound $Dy_2Ti_2O_7$. Here, the Dy^{3+} ions have a large magnetic moment (spin) and the Ti^{4+} ions are non-magnetic. The network structure of relevant oxygen ions is identical to the case of ice. Just like a proton in ice, there is a spin between the center oxygen and the corner oxygen, and the property of the spin is such that it can only point toward the center or corner. Because only two of the four spins around the center oxygen can point toward the center, ice-like frozen-in disorder is found in $Dy_2Ti_2O_7$. The same behavior is seen in $Ho_2Ti_2O_7$, and these compounds are called spin ice [13]. In 2008, theorists predicted that magnetic monopoles should be found in spin ice [14]. This was later confirmed by various experiments [15].

6.5 $KNiF_3$

Single crystals of $KNiF_3$ are shown in Fig. 6.5.

The study of second-order phase transitions has been one of the most fruitful fields of physics. Understanding such transitions in solids not only provides important insight into the properties of materials, but also offers new ideas to other fields of study such as elementary particle physics. The theory of second-order phase transitions has been developed mainly in the language of magnetism, because magnetic systems have the simplest set of microscopic interactions.

For second-order phase transitions, the behavior of the system as it approaches the transition is determined by three properties: (1) the dimensionality of the system, (2) the symmetry of the order parameter, and (3) whether the forces are of short or long range. As long as these properties are the same, different types of materials can undergo the same type of phase transition as magnetic systems (examples are the superfluid transition of liquid helium and atomic ordering in metallic alloys).

Although real materials have a three-dimensional crystal structure, the dimensionality of magnetic systems can be reduced to two or one if the interactions of spins are particularly strong along a plane or chain direction, respectively. The symmetry of the order parameter corresponds to whether spins can point only up or down (Ising), point anywhere along a plane (XY), or point anywhere along a space (Heisenberg). Both short- and long-range interactions are found in magnetic systems.

Transition-metal fluorides have served as important models for testing the theories of second-order phase transitions [17], because the spins in fluorides are well localized and fluorine ions are suitable for nuclear magnetic resonance and neutron

Fig. 6.5 $KNiF_3$. NiF_2, KF, $PbCl_2$, and NH_4HF_2 in 9.7:11.6:38.5:3.5 weight ratio were cooled from 900 to 350 °C at a rate of 4 °C per hour in a hand-sealed platinum crucible. Based on [16]

scattering measurements. The cubic perovskite $KNiF_3$ is a model three-dimensional Heisenberg system. K_2NiF_4, in turn, is a model two-dimensional Heisenberg system. $KCuF_3$ has the same atomic arrangements as $KNiF_3$, but the orbitals of the Cu^{2+} ions order in such a manner that $KCuF_3$ becomes a one-dimensional magnet. $CsNiF_3$ and $RbNiF_3$ are important hexagonal magnetic systems.

6.6 $KTiOPO_4$

Single crystals of $KTiOPO_4$ are shown in Fig. 6.6.

Since the first demonstration of laser action in 1960, many different types of lasers have been demonstrated. However, each type of laser typically generates only one or a few optical frequencies, and there are only a few lasers that have proven to be practical and commercially viable. Therefore, if there is a need to generate light at frequencies for which no convenient laser source is available, non-linear optical crystals are used to double or triple the frequency of laser light. There are various compounds of non-linear optical crystals on the market, each showing its own set of strengths and weaknesses. New and better crystals are being actively pursued by crystal growers in various laboratories [19].

$KTiOPO_4$, usually abbreviated as KTP, is one of the best non-linear optical crystals for changing near-infrared (1064 nm) light, from the Nd-doped $Y_3Al_5O_{12}$

Fig. 6.6 $KTiOPO_4$. K_2HPO_4, TiO_2, and WO_3 in 4:2:3 molar ratio were cooled from 1020 to 700 °C at a rate of 1 °C per hour in a covered platinum crucible. Based on [18]

garnet (YAG) laser, to green (532 nm) light. Its large optical non-linearity, high optical threshold, and outstanding thermal stability are especially important for high-power industrial and medical applications. Other well-known non-linear optical materials include KH_2PO_4, $LiNbO_3$, and BaB_2O_4.

Commercial crystals of KTP are usually grown by the top-seeded solution growth technique, using excess K_2O and P_2O_5 as a self-flux. High-quality crystals are also grown hydrothermally.

6.7 $La_{0.7}Pb_{0.3}MnO_3$

Single crystals of $La_{0.7}Pb_{0.3}MnO_3$ are shown in Fig. 6.7.

By the mid-1990s, a decade of feverish excitement concerning the high-temperature copper oxide superconductors was finally cooling down in the solid-state community. Many researchers were looking for a new topic, one not too

Fig. 6.7 $La_{0.7}Pb_{0.3}MnO_3$. La_2O_3, $MnCO_3$, PbO, and PbF_2 in 1.14:1.15:8.89:9.46 weight ratio were cooled from 1050 to 885 °C in a hand-sealed platinum crucible. The cooling rate was progressively increased from 0.2 to 1.2 °C per hour. Based on [20]

different from the copper oxides so that their acquired skills and knowledge could be put to some use. This is when colossal magnetoresistive manganese oxides appeared on the scene. Although these manganese oxides had been studied since the 1940s, new reports of large ("colossal") changes in resistivity under magnetic fields compelled many researchers to switch their focus from copper oxides to manganese oxides.

One of the prototypical compounds is perovskite $La_{1-x}Sr_xMnO_3$, where $x \sim 0.3$. Owing to the high tolerance of the perovskite structure, various rare-earth ions can replace La^{3+}, and other alkaline-earth ions or Pb^{2+} can substitute for Sr^{2+}. Each small change in composition seemed to affect the magnetic and electronic properties in a dramatic manner, providing an almost endless source of fascination to researchers. Other related phenomena, such as charge ordering and electronic phase separation, were also explored [21, 22], until the excitement began to fade in the 2000s.

6.8 $MgSiO_3$

Single crystals of $MgSiO_3$ are shown in Fig. 6.8.

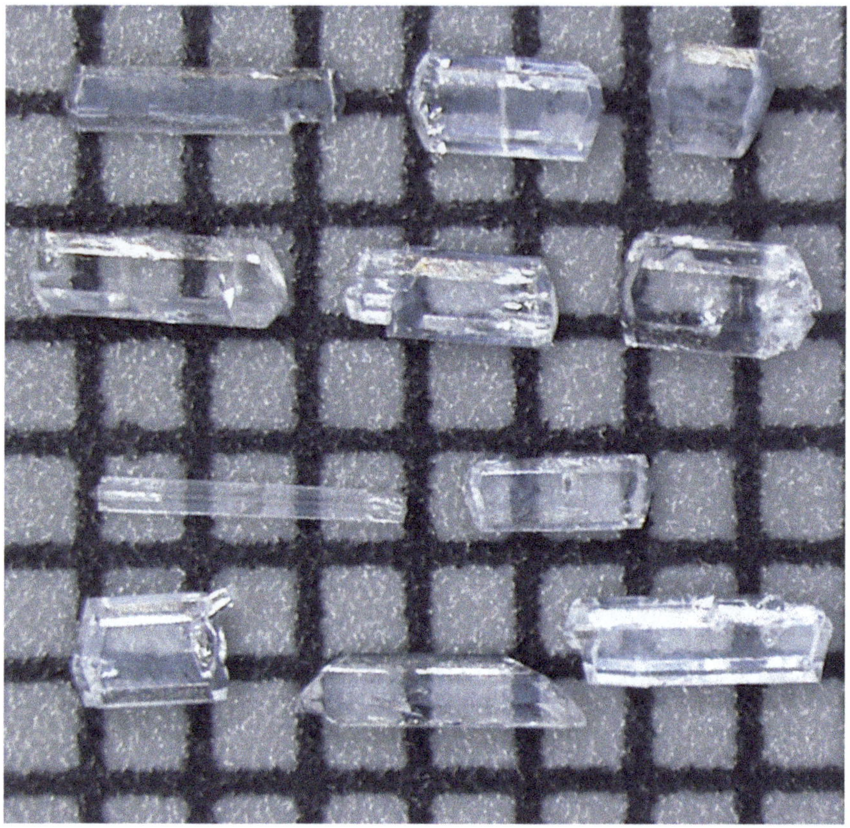

Fig. 6.8 $MgSiO_3$. MgO, SiO_2, MoO_3, $LiCO_3$, and V_2O_5 in 1.89:2.82:52.55:32.27:9.26 weight ratio were cooled from 950 to 875 °C at a rate of 0.44 °C per hour and then at a rate of 6.0 °C per hour to 600 °C in a covered platinum crucible. Based on [23]

The Earth has a radius of 6400 km and consists of three main distinct shells. The outermost shell is called the crust, and it extends down 5–70 km. Beneath the crust is the thick shell of the mantle, which extends down 2900 km. The innermost region is called the core and is made up mostly of iron. Both temperature and pressure increase with depth; it is estimated that the boundary between the mantle and core has conditions of about 2500–4000 K and 135 GPa.

The most abundant mineral at the lower part of the mantle is perovskite $MgSiO_3$, which is a high-pressure form of the mineral enstatite. Perovskite $MgSiO_3$ was discovered in 1975 [24], and it was believed to be the highest pressure phase for this composition; there was no known case of a perovskite compound transforming into another structure under increasing pressure.

Then, in 2004, scientists made a startling discovery in the laboratory: perovskite $MgSiO_3$ transforms into another structure above 125 GPa and 2500 K [25]. Dubbed "post-perovskite," this new structure has a layered structure unlike the nearly cubic perovskite structure. The stability condition of the post-perovskite phase implies that the lowest part of the mantle, from depths ~2700 to 2900 km, is made up of this new phase. This finding immediately explained the anomaly in the velocity of seismic waves at about ~2700 km, and also provided an important insight into the mechanisms of convection in the Earth's interior.

Single crystals of enstatite $MgSiO_3$ (the normal-pressure phase) and many other silicates can be grown using a flux of the Li_2O–MoO_3–V_2O_5 system. These crystals are often used by mineralogists and other geoscientists to study the properties of minerals and of the Earth.

6.9 $PbZn_{1/3}Nb_{2/3}O_3$

Single crystals of $PbZn_{1/3}Nb_{2/3}O_3$ are shown in Fig. 6.9.

Ferroelectrics exhibit a spontaneous electric polarization that can be switched in direction by applying an electric field. At the atomic scale, electric polarization comes from the displacement of positive and negative ions. Because of the interactions between polarization and lattice strain, ferroelectrics are also piezoelectrics—materials that convert electrical energy into mechanical energy, and mechanical energy into electrical energy. Piezoelectrics have many applications such as buzzers and tiny motors, and they are also used as transducers in medical ultrasound devices.

Fig. 6.9 $PbZn_{1/3}Nb_{2/3}O_3$. PbO, ZnO, Nb_2O_5, and B_2O_3 in 93.7:4.52:14.8:0.56 weight ratio were cooled from 1180 to 700 °C in a hand-sealed platinum crucible. The cooling rate was progressively increased from 2 °C per hour to 5 °C per hour. Based on [26]

Perovskite $PbTiO_3$ is a typical ferroelectric compound. It has a ferroelectric transition at 763 K, below which well-defined polarization is developed. On the other hand, isostructural $PbZn_{1/3}Nb_{2/3}O_3$ is a relaxor ferroelectric: it does not have a well-defined ferroelectric transition, and heterogeneous polarization at the nanometer scale is found even in single crystals. This behavior arises from the random distribution of Zn^{2+} and Nb^{5+} ions within the same structural site, which creates a strong random electric field. Relaxor compounds such as $PbZn_{1/3}Nb_{2/3}O_3$ and $PbMg_{1/3}Nb_{2/3}O_3$ show a number of unusual physical properties [27].

Relaxor compounds are also excellent piezoelectric materials. One of the best piezoelectric properties is found in solid solutions between $PbZn_{1/3}Nb_{2/3}O_3$ and $PbTiO_3$, near the composition where the crystal structure undergoes transformation. Excellent properties mean that piezoelectric devices can become smaller and more reliable, which is especially important in medical and military applications.

6.10 SmB_6

Single crystals of SmB_6 are shown in Fig. 6.10.

SmB_6 provides an excellent account of how some compounds never get old in solid-state physics. Its peculiar properties first became known in 1969, when researchers reported that SmB_6 changes from a metal to an insulator with decreasing temperature [29]. This behavior was ascribed to a change in the valence of the Sm ions, a phenomenon that became popular in the 1970s and 1980s within the field of valence instabilities and valence fluctuations in solids.

Studies on SmB_6 continued throughout the 1990s, this time as an odd member of heavy fermion compounds. Heavy fermions arise from the interaction between conduction electrons and localized magnetic moments in rare-earth and actinide compounds. Unlike many other heavy fermions, the important hybridized band in SmB_6 is completely filled, making it an insulator at low temperatures. The name "Kondo insulator" is used to describe this phenomenon [30].

SmB_6 was believed to be fairly well understood, except for one fact: it becomes metallic once again at a very low temperature. Although researchers tried to brush aside this problem by invoking defects and impurities, such ideas were not consistent with other properties.

When topological insulators—solids that conduct electricity like a metal only across their surfaces—became a new field of study in the late 2000s [31], physicists realized that the peculiar properties of SmB_6 could be explained in this context [32]. As topological insulators are poised to change the way we view solids, studies on SmB_6 will continue into the future.

Fig. 6.10 SmB_6. Sm, B, and Al in SmB_6:Al = 1:20 weight ratio were cooled from 1300 to 600 °C over a period of 100 h in an alumina crucible under flowing argon. The flux was dissolved in dilute hydrochloric acid. Based on [28]

6.11 $TbMn_2O_5$

Single crystals of $TbMn_2O_5$ are shown in Fig. 6.11.

Ferroelectrics have spontaneous electric polarization which can be switched by an applied electric field. Ferroelectricity is found in compounds where the positive and negative ions are displaced in such a way that the center of symmetry is lost. In almost all cases, ferroelectric displacement requires the ionic shells to be either empty or completely filled with electrons.

On the other hand, magnetic order is related to the ordering of spins. The spins come from electrons in incomplete ionic shells.

These mutually exclusive requirements explain why there are only a few compounds that show both ferroelectricity and magnetic order [34]. This is unfortunate, because the combination of these two states can lead to many interesting phenomena with possible practical applications.

The study of such rare compounds, called ferroelectric magnets or multiferroics, received a huge boost in 2003. In that year, physicists found that the magnetic order in the perovskite $TbMnO_3$ is accompanied by ferroelectric order, and that electric polarization can be switched by a magnetic field [35]. Soon afterward, theorists showed that magnetic order without the center of symmetry can induce electric

Fig. 6.11 $TbMn_2O_5$. Tb_4O_7, $MnCO_3$, B_2O_3, PbO, PbF_2, and PbO_2 in 9.2:11.5:1:36:54:3.3 weight ratio were cooled from 1280 to 950 °C at a rate of 1.2 °C per hour in a hand-sealed platinum crucible. Based on [33]

polarization, even if the atomic structure has a center of symmetry [36]. This realization urged researchers to re-examine known compounds with such magnetic order, and many new multiferroics were soon discovered.

$TbMn_2O_5$ was known for a long time to show a number of complex magnetic and ferroelectric transitions. The report of its electric polarization reversal under an applied magnetic field in 2004 [37] launched systematic studies of this compound, as well as of other members of RMn_2O_5, where R is a rare earth.

6.12 VO_2

Single crystals of VO_2 are shown in Fig. 6.12.

Metal–insulator transitions provide important clues for understanding the electronic interactions in solids [39]. Determining the origin and mechanism of the metal–insulator transition is difficult, however, because it can be electronically, structurally, or magnetically driven. In some cases, all three factors contribute to the transition.

On cooling, the resistivity of VO_2 jumps by four orders of magnitude at 340 K, indicating a sharp metal–insulator transition. There is a strong lattice distortion at the transition, where two adjacent V^{4+} ions combine to form one pair. Ever since

Fig. 6.12 VO_2. V_2O_5 in a silica glass crucible was heated at 1000 °C for a period of 156 h under flowing argon. The flux was dissolved in dilute ammonia. Based on [38]

this transition was reported in 1959 [40], the main question has been whether it is a consequence of strong electron–electron interactions or electron–lattice interactions. Recent studies provide important insight into this matter [41, 42], but the question is far from settled.

The flux growth of VO_2 utilizes the fact that the melting temperature of V_2O_5 at 690 °C is much lower than that of VO_2 at 1700 °C—if V_2O_5 is kept above 690 °C under a reduced oxygen partial pressure, VO_2 becomes the stable phase and crystallizes out from the liquid.

References

1. N. Ni, M.E. Tillman, J.-Q. Yan, A. Kracher, S.T. Hannahs, S.L. Bud'ko, P.C. Canfield, Phys. Rev. B **78**, 214515 (2008)
2. A.P. Mackenzie, Y. Maeno, Rev. Mod. Phys. **75**, 657–712 (2003)
3. Y. Kamihara, T. Watanabe, M. Hirano, H. Hosono, J. Am. Chem. Soc. **130**, 3296–3297 (2008)
4. J. Paglione, R.L. Greene, Nat. Phys. **6**, 645–658 (2010)
5. G.H. Larson, A.W. Sleight, Phys. Lett. **28A**, 203–204 (1968)
6. R.P. van Stapele, in *Ferromagnetic Materials*, vol. 3, ed. by E.P. Wohlfarth, (North-Holland, Amsterdam, 1982), pp. 603–745
7. K.G. Barraclough, Prog. Cryst. Growth Charact. **1**, 57–84 (1977)
8. E.I. Speranskaya, Inorg. Mater. (USSR) **3**, 1271–1277 (1967)
9. P. Lemmens, G. Güntherodt, C. Gros, Phys. Rep. **35**, 1–103 (2003)

10. M. Hase, I. Terasaki, K. Uchinokura, Phys. Rev. Lett. **70**, 3651–3654 (1993)
11. B.M. Wanklyn, A. Maqsood, J. Mater. Sci. **14**, 1975–1981 (1979)
12. L. Pauling, J. Am. Chem. Soc. **57**, 2680–2684 (1935)
13. S.T. Bramwell, M.J.P. Gingras, Science **294**, 1495–1501 (2001)
14. C. Castelnovo, R. Moessner, S.L. Sondhi, Nature **451**, 42–45 (2008)
15. C. Castelnovo, R. Moessner, S.L. Sondhi, Annu. Rev. Condens. Matter Phys. **3**, 35–55 (2012)
16. M. Safa, B.K. Tanner, B.J. Garrard, B.M. Wanklyn, J. Cryst. Growth **39**, 243–249 (1977)
17. L.J. de Jongh, A.R. Miedema, Adv. Phys. **23**, 1–260 (1974)
18. A.A. Ballman, H. Brown, D.H. Olson, C.E. Rice, J. Cryst. Growth **75**, 390–394 (1986)
19. P. Thomas, Phys. World **3**(3), 34–38 (1990)
20. N. Ghosh, S. Elizabeth, H.L. Bhat, U.K. Rößler, K. Nenkov, S. Rößler, K. Dörr, K.-H. Müller, Phys. Rev. B **70**, 184436 (2004)
21. M.B. Salamon, M. Jaime, Rev. Mod. Phys. **73**, 583–628 (2001)
22. Y. Tokura, Rep. Prog. Phys. **69**, 797–851 (2006)
23. T. Tanaka, H. Takei, J. Cryst. Growth **180**, 206–211 (1997)
24. T. Liu, Geophys. Res. Lett. **2**, 417–419 (1975)
25. M. Murakami, K. Hirose, K. Kawamura, N. Sata, Y. Ohishi, Science **304**, 855–858 (2004)
26. L. Zhang, M. Dong, Z.-G. Ye, Mater. Sci. Eng., B **78**, 96–104 (2000)
27. A.A. Bokov, Z.-G. Ye, J. Mater. Sci. **41**, 31–52 (2006)
28. V.N. Gurin, M.M. Korsukova, S.P. Nikanorov, I.A. Smirnov, N.N. Stepanov, S.G. Shul'man, J. Less-Common Met. **67**, 115–123 (1979)
29. A. Menth, E. Buehler, T.H. Geballe, Phys. Rev. Lett. **22**, 295–297 (1969)
30. G. Aeppli, Z. Fisk, Comments Condens. Matter Phys. **16**, 155–165 (1992)
31. J.E. Moore, Nature **464**, 194–198 (2010)
32. M. Dzero, K. Sun, V. Galitski, P. Coleman, Phys. Rev. Lett. **104**, 106408 (2010)
33. B.M. Wanklyn, J. Mater. Sci. **7**, 813–821 (1972)
34. N.A. Hill, J. Phys. Chem. B **104**, 6694–6709 (2000)
35. T. Kimura, T. Goto, H. Shintani, K. Ishizaka, T. Arima, Y. Tokura, Nature **426**, 55–58 (2003)
36. S.-W. Cheong, M. Mostovoy, Nat. Mater. **6**, 13–20 (2007)
37. N. Hur, S. Park, P.A. Sharma, J.S. Ahn, S. Guha, S.-W. Cheong, Nature **429**, 392–395 (2004)
38. H. Sasaki, A. Watanabe, J. Phys. Soc. Jpn. **19**, 1748 (1964)
39. M. Imada, A. Fujimori, Y. Tokura, Rev. Mod. Phys. **70**, 1039–1263 (1998)
40. F.J. Morin, Phys. Rev. Lett. **3**, 34–36 (1959)
41. J.H. Park, J.M. Coy, T.S. Kasirga, C. Huang, Z. Fei, S. Hunter, D.H. Cobden, Nature **500**, 431–434 (2013)
42. J.D. Budai, J. Hong, M.E. Manley, E.D. Specht, C.W. Li, J.Z. Tischler, D.L. Abernathy, A.H. Said, B.M. Leu, L.A. Boatner, R.J. McQueeney, O. Delaire, Nature **515**, 535–539 (2014)

Appendix
Flux-Grown Crystals Published in *Journal of Crystal Growth* since 1975

A comprehensive list of flux-grown crystals up to 1975 is given in *Crystal Growth from High-Temperature Solutions* by D. Elwell and H.J. Scheel (Academic Press, London, 1975). To provide some more recent examples, this Appendix lists the flux-grown crystals that have been reported in *Journal of Crystal Growth* from 1975 to 2016 inclusive. Although efforts were made to include as many relevant papers as possible, some important studies may have been overlooked. Evaporation growths are included, but seeded growths are not, and only one example for each compound is shown in most cases (the presence of other studies is remarked upon). High-pressure growths and growths that result in crystals much smaller than 1 mm are not included. The list provides the highest temperature used in the experiment (such as the starting temperature of slow cooling), the maximum crystal dimension in millimeters, the published volume, and the first page number. The published year is not shown, but the following examples should give approximate times of publication: Vol. 30 (1975); 50 (1980); 75 (1986); 100 (1990); 150 (1995); 200 (1999); 250 (2003); 300 (2007); 350 (2012); 400 (2014). In few instances the paper is published in multi-volume proceedings; in such a case the correct volume is given in the remarks column. Because of their sheer variety, high-temperature copper oxide superconductors are not included in the list.

M. Tachibana, *Beginner's Guide to Flux Crystal Growth*, NIMS Monographs,
DOI 10.1007/978-4-431-56587-1

Flux-grown crystals published in *Journal of Crystal Growth* from 1975 to 2016

Crystal	Flux[a]	T_{max} (°C)	mm	Remarks	Vol.	Page
$A_{0.3}MoO_3$	$A_2O–MoO_3$	605	8	*A* = K, Rb, Cs	70	476
$A_{0.9}Mo_6O_{17}$	$A_2O–MoO_3$	572	5	*A* = Na, K	70	476
$A_2Nb_4O_{11}$	A_2O	1200	15	*A* = Cs, Rb; Vol. 237–239	237	703
A_2SiO_4	$PbO–PbF_2$	1270	10	*A* = Co, Zr, Th, Zr, Zn, Mg; also Si=Ge	37	51
A_2SnO_4	$Bi_2O_3–V_2O_5$	1300	3	*A* = Mg, An, Co	59	662
ABO_3	$PbO–PbF_2$	1000	8	*A* = Fe, Ga, In, Sc, Lu	455	55
Ag(Ta, Nb)O_3	$Ag_2O–V_2O_5$	1152	5	Also previous study	96	703
Al_2O_3:Cr	$Bi_2O_3–PbF_2–La_2O_3$	1270		Also other fluxes	280	551
$Al_{71}Pd_{21}Mn_8$	Al	875		Quasicrystal	225	155
$Al_{80}Ni_{11}Co_{17}$	Al	1200		Quasicrystal	225	155
Al–Mg–B	Al	1500	4	Also other borides	99	998
AlN	Fe	1700	1		34	263
ANb_2O_6	$Na_2B_4O_7$	1240	8	*A* = Mg, Zn, Ba	58	463
APd_3O_4	KOH	750	2	*A* = Ca, Sr	216	299
$AR_9(SiO_4)_6O_2$	(Na, Li)MoO_4	1380	4	*A* = Li, Na; *R* = Eu, Nd	99	879
ASb_2O_6	$V_2O_5–B_2O_3$	1000	2.5	*A* = Mn, Co, Ni, Cu	154	334
(Ba, Ca)TiO_3	KF	1160	10		94	125
(Ba, K)Fe_2As_2	Sn			Solubility study	316	85
(Ba, Sr)$_2Zn_2Fe_{12}O_{22}$	$Na_2O–Fe_2O_3$	1420	27		83	403
(Ba, Sr)TiO_3:Co	KF	1220	5	Vol.: 237–239	237	858
$Ba_2Fe_{10}Sn_2CoO_{22}$	$BaO–B_2O_3$	1260	3	Also other compositions	110	617
Ba_2Ho(Ru, Cu)O_6	$PbO–PbF_2$	1250	3		290	490
$BaAlBO_3F_2$	NaF		9.5		260	287

(continued)

(continued)

Crystal	Flux[a]	T_{max} (°C)	mm	Remarks	Vol.	Page
BaB_2O_4	$NaCl–Na_2O$	880		Other studies exist	97	613
$BaCoO_3$	NaOH–KOH	750	2		430	52
$Ba(Fe, Co)_2As_2$	(Fe, Co)As	1190	10	Other studies exist	321	55
$BaFe_{12}O_{19}$	$Na_2O–B_2O_3$	1200	10		169	509
$BaFe_2(As, P)_2$	BaAs–BaP	1150	5		446	39
$BaMoO_4$	LiCl	700	3		53	627
$Ba(Pb, Bi)O_3$	$PbO–PbO_2–Bi_2O_3$	1080	6	Also other studies	151	295
$BaSO_4$	LiCl–KCl	600	15		234	533
$BaTiO_3$	KF	1200	7	Published in 2017	468	753
$BaTiO_3$:F	$LiF–BaF_2–LiBO_2$	1130	7		67	79
$Be_3Al_2Si_6O_{18}$:Cr	$V_2O_5–Li_2O–P_2O_5$	1000			193	648
BeO	$K_2MoO_4–MoO_3$	1100			42	284
$Bi_2Ru_2O_7$	$Bi_2O_3–V_2O_5$	1150	1		68	647
Bi_2TeI	Bi	850	5		440	26
$Bi_2Ti_2O_7$	$Bi_2O_3–V_2O_5$	1300	10	Also other Bi–Ti–O phases	41	317
Bi_2WO_6	$WO_3–Na_2O–NaF$	900	10	Also other studies	54	217
$Bi_4Ti_3O_{12}$:Nd	Bi_2O_3	1250	10	Other studies exist	310	2471
$Bi_6Mo_2O_{15}$	Various		6	Other phases studied	51	377
$BiFeO_3$	Bi_2O_3	852	2		318	936
$BiFeO_3$	$Bi_2O_3–B_2O_3$	850	5		129	515
$BiFeO_3–PbTiO_3$	$PbO–Bi_2O_3$	1200	5		285	156
$Bi(Sc,Ga)O_3–PbTiO_3$	$Pb_3O_4–Bi_2O_3$	1250	8		247	131
$BiScO_3–PbTiO_3$	$Pb_3O_4–Bi_2O_3$	1200	15		236	210
BN	Ni–Cr	1500	2	Also other studies	403	110

(continued)

(continued)

Crystal	Flux[a]	T_{max} (°C)	mm	Remarks	Vol.	Page
BP	Ni_2P	1200	5		33	53
$Ca_2Al_2SiO_7$	$PbO–PbF_2$	1200	3	Other related compounds	102	919
$(Ca_2CoO_3)_{0.62}(CoO_2)$	$SrCl_2$	927	5	Sr-doped	276	519
Ca_2GeO_4:Cr	$CaCl_2–CaF_2$	1050	10		211	295
$Ca_3Co_4O_9$	K_2CO_3	895	10	Other fluxes studied	277	246
$Ca_3Si_2O_7 \cdot 1/3CaCl_2$	$CaCl_2$	1300	15		52	660
Ca_4PtO_6	$CaCl_2–CaF_2$	1000	2		51	1
$CaCu_3Ti_4O_{12}$	$CuO–TiO_2$	1200	6		408	60
CaF_2	KCl	1000	10	Also NaCl, $CaCl_2$ fluxes	44	625
Ca(Fe, Co)AsF	CaAs	1230	1		451	161
CaO	$LiF–CaF_2$	1250	3		39	223
$CaTiO_3$	KF	1170	2	Also other fluxes	94	125
CaV_3O_7	$NaVO_3$	900	8	CaV_4O_9 also grown	240	170
$CdCr_2O_4$	$Bi_2O_3–V_2O_5$	1210	2.5	Also PbO flux	54	607
$CdCr_2Se_4$	$CdCl_2$	900	4		40	253
$CdGa_2O_4$	$PbO–B_2O_3$	1250	1.2		171	131
$CdGeAs_2$	Bi	750	20		28	138
$CdGeP_2$	Bi	830	12		50	567
$CdTiO_3$	KF	1120	2		94	125
$(Ce, Pr)O_2$	$Na_2B_4O_7–NaF$	1200	3	Also PbF_2-based flux	87	463
$CeFe_2$	Ce	1100		Also other rare earths	225	155
CeO_2	$PbO–PbF_2$	1330	8	Additives used	66	346
CeO_2:Yb	PbF_2	1240	3		35	239
CeRuPO	Sn	1500	2.5		310	1875

(continued)

(continued)

Crystal	Flux[a]	T_{max} (°C)	mm	Remarks	Vol.	Page
$(Co, Mn)_3O_4$	PbF_2	1170	4	Evaporation	344	65
$Co_3(Sn, In)_2S_2$	Sn–In–Pb	1050	7		426	208
$CoFe_2O_4$	$Na_2B_4O_7$	1350	5		289	605
$CoFe_2O_4$:Dy	$Na_2B_4O_7$	1350			340	171
CoV_2O_6	V_2O_5–$PbCl_2$	800	10		388	103
Cr_2O_3	$K_2Cr_2O_7$	1200	7	Other studies and fluxes	47	551
$CsAu_2X_6$	Cs*X*–Au	630	0.6	X = Br, I	355	13
$CsNiMF_6$	CsCl–NH_4HF_2	1200	12	M = Fe, Cr	54	610
$CsVF_4$	$PbCl_2$–NH_4HF_2	860	12	Also other fluorides	47	159
$CuAlO_2$	Cu_2O	1160	2		310	4325
$CuGaS_2$	Pb		6	Also Sn flux	53	451
$CuGaSe_2$	In	1000		Also other studies	84	673
$CuGeO_3$	Bi_2O_3	1050	40	Co, Ga doped; other studies	204	311
$Cu(In, Ga)S_2$	CsCl	1030	4		412	16
$CuIr_2S_4$	Bi	1000	1		210	772
CuO	BaO–B_2O_3	1000	35	$BaCuO_2$ also grown	129	239
$CuSb_2O_6$	V_2O_5	1000	2		72	753
CuV_2O_6	V_2O_5–K_2SO_4	600	3		231	498
$CuWO_4$	Na_2WO_4–$CuCl_2$	850	40		88	379
$DyFeO_3$	PbO–PbF_2–B_2O_3	1245	20	MoO_3 added	29	281
$Er_2Si_2O_7$	PbO–PbF_2–MoO_3	1270	3	Also KF added	43	336
Eu_2O_3	NaF	1200	5		41	309
$(Fe, Ga)_2O_3$	Bi_2O_3–Na_2O	1250			87	578
Fe_2O_3	PbO–V_2O_5	1250	8	Solubility study	49	182

(continued)

(continued)

Crystal	Flux[a]	T_{max} (°C)	mm	Remarks	Vol.	Page
Fe_2O_3	V_2O_5–KCl	900	5	37 fluxes tried	58	636
$FeBO_3$	PbO–PbF_2	870	18		71	607
FeS_2	Te	650	7	Temp. gradient	92	287
$FeSi_2$	Zn	930	1	Vol.: 237–239	237	1971
Ga_2GeO_4	Li_2O–B_2O_3			Other phases studied	426	25
$GaPO_4$	Li_2O–MoO_3	950	8	Other studies exist	310	1455
Gd_2GeMoO_8:Yb	MoO_3	1100	4		318	991
$GdRh_2Si_2$	In	1550			419	37
In_5S_4	Sn	1100	2		52	673
$InBO_3$	PbO	1300	65	Part of seeded study	64	385
$InBO_3$:Tb	$LiBO_2$	1150	30		99	799
(K, Li)TaO_3	K_2O	1325			56	673
(K, Na)NbO_3	NaF_2	1250			46	274
$K_{1.98}Fe_{1.98}Sn_{6.02}O_{16}$	K_2O–MoO_3–B_2O_3	1000	3		390	88
$K_{1+x}Fe_{11}O_{17}$	B_2O_3–K_2O–KF	1200	6.6		71	253
$K_5V_3F_{14}$	$PbCl_2$	850	6	Also $KTiF_3$, VF_2	33	165
$KBe_2(BO_3)F_2$	KF–B_2O_3	800	30	Also other studies	318	610
$KFeF_3$	$PbCl_2$–NH_4HF_2	820	3	Also $RbFeF_3$, CrF_2, KVF_4	29	301
KMF_3	$PbCl_2$–NH_4HF_2	960	5	M = Fe, Co, Ni	39	243
$KNbB_2O_6$	$K_2B_4O_7$	975	5	Also other compositions	220	263
$KNiF_3$	KCl–KHF_2	1050	8		54	610
K(Ta, Nb)O_3	K_2O	1400	35	Also other studies	59	468
$KTiOPO_4$	K_2WO_4–P_2O_5	1000	10	Other studies/isomorphs	75	390
(La, Na)Fe_2As_2	NaAs	1150	2	Reaction w/ Al_2O_3 crucible	416	62

(continued)

(continued)

Crystal	Flux[a]	T_{max} (°C)	mm	Remarks	Vol.	Page
(La, Pr)AlO_3	$PbO–PbF_2–B_2O_3$	1300	15	MoO_3 also added	33	150
$La_{2/3}TiO_{3-x}$	$KF–Na_2B_4O_7$	1000	3	Different phase at 950 °C	96	490
$La_2Cu_2O_5$	CuO	1115	7		212	142
La_5Pb_3O	Co	1150	4	Reaction w/ Al_2O_3 crucible	416	62
$LaAlO_3$:Cr	PbF_2	1270	5		47	315
$LaBO_3$	$PbO–B_2O_3$	1240		Solubility study	58	111
LaCuOS	NaCl–KCl	850	3		311	114
Li_2MnO_3	LiCl	900	4		66	257
$Li_3Ba_2Nd_3(MoO_4)_8$	Li_2MoO_4	1000	40		381	61
Li_3ThF_7	$LiCl–ZnCl_2$	900	4.7	Also other compounds	40	157
$Li_4Ti_5O_{12}$	$LiF–LiBO_2$	1000	2	Also $LiTi_2O_4$	250	139
$LiAlSiO_4$	$LiF–AlF_3$	1100	15		42	289
$LiCuVO_4$	$LiVO_3–LiCl$	560	4	Also other studies	220	345
$LiInGeO_4$:Cr	Bi_2O_3	1070	20		274	149
$LiMo_3Ni_2O_{12}$	$Li_2O–MoO_3$	1300	8	Also Ni=Mg	34	301
$LiNd(MoO_4)_2$	$Li_2O–MoO_3$	700	2		423	1
(Mg, Fe)SiO_3	$Li_2O–MoO_3–V_2O_5$	1014	8		200	155
$MgAl_2O_4$	$PbO–PbF_2–B_2O_3$	1270		Intentional twinning	49	753
$MgSiO_3$	$Li_2O–MoO_3–V_2O_5$	950	8	Also other studies	180	206
$MgSiO_3$:Ti/Ni	$Li_2O–MoO_3–V_2O_5$	950	1.5		329	86
$Mn_2V_2O_7$	SrV_2O_6	1080	5		310	171
MnSi	Ga	1200	2.4		229	532
$MnSiO_3$	$MnCl_2$	920	7		94	981
(Mo, W)Se	Sb	1100	10	Also Se, Te, Bi, $PbCl_2$ flux	76	93

(continued)

(continued)

Crystal	Flux[a]	T_{max} (°C)	mm	Remarks	Vol.	Page
MoO_3	Na_2MoO_4	670	30		194	195
$(Na, K)_{1/2}Bi_{1/2}TiO_3$	Bi_2O_3–Na_2O–K_2O	1100	5		281	364
$(Na, K)Fe_xSe_2$	NaCl	720			405	1
$Na_{1/2}Bi_{1/2}TiO_3$–$BaTiO_3$	Bi_2O_3	1300	13	Also other studies	441	64
$Na_2Nd_2Pb_6(PO_4)_6Cl_2$	PbO–$PbCl_2$	1000	5		43	81
$Na_2Ti_2Sb_2O$	NaSb	1100	4.2		265	571
$Na_2W_4O_{13}$	Na2O	1100	35		229	477
$Na_5Fe_3F_{14}$	NaCl–$CoCl_2$	650	2	Also Fe=Mn, Ni	32	211
Na_8Si_{46}	Na–Sn	450	1.5	Evaporation	450	164
$NaCa_2M_2V_3O_{12}$	Na_2O–V_2O_5–PbO	1100	10	M = Mg, Co, Mn	52	650
$NaCu_2O_2$	CuO	940	7		263	338
$Na(V, Ti)_2O_5$	$NaVO_3$	800	12		210	646
NaV_2O_5	$NaVO_3$	800	10		181	314
$Na_xCo_2O_4$	NaOH–NaCl	550	5		310	665
$Na_xTi_4O_8$	$NaBO_3$	1265	14	Also Ti=Fe, Ni	43	153
Nb_5Sn_2Ga	Sn–Ga	1400	10	Also Ta_5SnGa_2, $V_5Sn_5Ga_3$	99	969
NbC	Ni	1400	2		62	557
$(Nd, Dy)Fe_3(BO_3)_4$	Bi_2O_3–B_2O_3–MoO_3	1000	2		312	2427
$(Nd, La)P_5O_{12}$	H_3PO_4	700	10	Also other studies	35	329
$(Nd, Pr)GaO_3$	PbO–PbF_2–MoO_3	1280	4.5		128	699
Nd_3BWO_9:Yb	PbO	1100	10		247	467
$Nd_4Ca_2Ti_6O_{20}$	PbO	1280	10	Other fluxes studied	65	576
NdOCl	$NdCl_3$	1350	10		57	194
$NdTa_7O_{19}$	$Li_2B_4O_7$		3		224	67

(continued)

(continued)

Crystal	Flux[a]	T_{max} (°C)	mm	Remarks	Vol.	Page
Ni_2SiO_4	$Li_2O–MoO_3$	1400	10		33	193
$Ni_3V_2O_8$	$SrO–V_2O_5$	1000	3		297	1
$NpPd_3$	Pb	1050	3		320	52
(Pb, La)(Z, Sn, Ti)O_3	$PbO–PbF_2–B_2O_3$	1200	2	Also other studies	318	860
$Pb_2Ru_2O_{6.5}$	PbO	1250	9		271	445
$PbB'_{1/2}B''_{1/2}O_3$	$PbO–PbF_2–B_2O_3$	1200	6	B′B″ = InNb, InTa, YbNb, YbTa, MgW	310	2767
$PbCo_{1/2}W_{1/2}O_3$	PbO	1230	17	Also other studies	82	396
$PbFe_{1/2}Nb_{1/2}O_3$	PbO	1260	15		56	541
$PbFe_{1/2}Ta_{1/2}O_3$	PbO	1230	5		82	396
$PbFe_{1/2}W_{1/2}O_3$	$PbO–B_2O_3$	1030	3		167	628
$PbIn_{1/2}Nb_{1/2}O_3–PbTiO_3$	$PbO–PbF_2–B_2O_3$	1200	20		229	299
$PbMg_{1/3}Nb_{2/3}O_3–PbTiO_3$	$PbO–Pb_3O_4–B_2O_3$	1090	6	Other studies exist	289	134
$PbMg_{1/3}Nb_{2/3}O_3–PbTiO_3$	$PbO–H_3BO_3$	1070	14	$BiZn_{1/2}Ti_{1/2}O_3$ doped	318	839
$PbMg_{1/3}Ta_{2/3}O_3–PbTiO_3$	$PbO–Pb_3O_4–B_2O_3$	1130	4	Also previous study	310	594
$PbMn_{1/2}Nb_{1/2}O_3$	PbO	1260	6		56	541
$PbSc_{1/2}Nb_{1/2}O_3–PbTiO_3$	$PbO–B_2O_3$	1300	4	Other studies exist	250	118
$PbTiO_3$	PbO	1100	5	Other studies exist	128	867
$PbWO_4$	Na_2WO_4	950	8		57	452
$PbYb_{1/2}Nb_{1/2}O_3–PbTiO_3$	Pb_3O_4	1200	6		234	415
$PbZn_{1/3}Nb_{2/3}O_3–PbTiO_3$	PbO	1250	30	Other studies exist	216	311
Pb(Zr, Ti)O_3	PbO	1170	10		33	29
Pd(Co, Mg)O_2	$PdCl_2$	700	1	Other studies exist	226	277
$PdCrO_2$	NaCl	880	3.5		312	3461
(*R*, Pb)MnO_3	$PbO–PbF_2$	1050	4	*R* = La, Nd, Pr; other studies	275	e163

(continued)

(continued)

Crystal	Flux[a]	T_{max} (°C)	mm	Remarks	Vol.	Page
$R_{2/3}TiO_{3-x}$	$KF–Na_2B_4O_7$	950	4	R = Nd, Sm, Gd	99	875
$R_2Ge_2O_7$	$PbO–PbF_2–MoO_3$	1270		Also K_2O–KF added	43	336
R_2GeMoO_8	$PbO–PbF_2–MoO_3$	1290			43	336
R_2MGa_{12}	Ga	1150	5	R = Pr, Nd, Sm; M = Ni, Cu	312	1098
$R_2Si_2O_7$	$PbO–PbF_2–MoO_3$	1270	14	R = rare earth	46	671
$R_2Sn_2O_7$	$Na_2B_4O_7$–NaF	1000	2	R = rare earth; a 2017 paper	468	335
$R_2Ti_2O_7$	$KF–Na_2B_4O_7$	1000	3	R = Dy, Er, Yb	99	875
$R_3Al_5O_{12}$	$PbO–PbF_2–B_2O_3$	1285	8	Also Al=Ga	54	610
$RAl_3(BO_3)_4$	$K_2SO_4–MoO_3–B_2O_3$	1140	5	Other studies exist; R = Nd	89	295
$RAlO_3$	$PbO–PbF_2–B_2O_3$	1295	5	MoO_3 added	29	281
RB_6	Al	1500	1	R = La, Eu, Y, Ce, Ba, Cs	44	287
RBO_3	$PbO–PbF_2$	1330	12	Other studies exist	54	610
R–B–Si	Cu–Si	1650	5	Small R–B–C(N) grown	271	159
$RGaO_3$	$PbO–PbF_2–MoO_3$	1260	4	R = La, Pr, Nd; B_2O_3 added	94	125
$RKMo_2O_8$	$K_2O–MoO_3$	1270		Also R_2MoO_6, R_6MoO_{12}	43	93
$RKMo_2O_8$	$PbO–MoO_3$	1290			43	336
R–Mg–Zn	Mg–Zn	700	6	Quasicrystals	225	155
RMn_2Si_2	Pb	1350	2.5	Other studies exist	244	267
RNi_2B_2C	Ni_2B	1500		R = rare earth	225	155
RPO_4	$PbO–P_2O_5$	1330	4	Also RVO_4 with V_2O_5 flux	43	336
RPO_4	$PbO–P_2O_5$	1350	15	R = rare earth	63	77
RRh_3B_2	Cu	1350	3	R = Gd, Er, Tm	229	521
RT_2Ge_2	T–Ge	1190		R = rare earth; T = Ni, Cu	225	155
RuX_2	Bi	1000	5	X = S, Se, Te	83	517

(continued)

(continued)

Crystal	Flux[a]	T_{max} (°C)	mm	Remarks	Vol.	Page
RVO_4	$PbO–V_2O_5$	1360	30	Other studies exist	79	534
Sc_2O_3	$PbO–PbF_2–V_2O_5$	1330	2		104	672
$Sc_2(WO_4)_3$	$Bi_2O_3–WO_3$	1200	4		143	362
Si	Na	900	3	Evaporation	355	109
Si_3N_4	Si	1600			46	143
SnO	Cu_2O	1320			233	259
$(Sr, Pb)(Cr, Ga)_{12}O_{19}$	$PbO–PbF_2–B_2O_3$	1360	4		165	179
Sr_2NiWO_6	$SrCl_2$	1100	1		421	39
Sr_3NiPtO_6	K_2CO_3	1150		Other related compounds	204	122
Sr_4PtO_6	$SrCl_2$	1150	4		64	395
$SrCu_2(BO_3)_2$	$Na_2B_4O_7$	900	7	Also previous study	277	541
$SrGa_{12}O_{19}$	Bi_2O_3	1350	15		61	284
$SrNdFeO_4$	Bi_2O_3		15	Also other Sr–Nd–Fe–O	32	332
$SrRFeO_5$	PbO–SrO	1550	2	Also Fe=Al	47	739
$SrZrO_3$	KF–NaF–LiF	1200	0.1		94	125
TaC	Ni	1800	1.5	Also Ni–Co flux	75	454
$Th_{0.5}Pb_{0.5}VO_4$	$PbO–V_2O_5$	1300		ThO_2 also grown	71	289
$ThGeO_4$	$PbO–PbF_2–MoO_3$		10	Also K_2O–KF added	43	336
ThO_2	$PbO–PbF_2$	1330	8	Additives used	66	346
$TiAs_2$	Cs_3As_7	900	2	Also other compounds	217	250
TiB_2	Al	1550	5	Other B- and C-compounds	33	207
$TmVO_4$	$LiVO_3$	1150	7	Vol.: 198/199	198	449
U_3Bi_4	Bi	1080			172	459
$V(PO_3)_3$	H_3PO_4	450	9		63	209

(continued)

(continued)

Crystal	Flux[a]	T_{max} (°C)	mm	Remarks	Vol.	Page
WO_3	PbF_2	1250	25	Other fluxes studied	88	143
$Y_2Cu_2O_5$	Cu_2O–CuO	1350	3	Small rods	141	153
$Y_2Cu_2O_5$	Cu_2O	1200	1		141	150
$Y_3Fe_5O_{12}$	PbO–B_2O_3	1250	10	Other studies exist	28	231
$YAlO_3$	PbO–PbF_2	1280	2	$Y_3Al_5O_{12}$ also obtained	36	255
$Yb_{0.24}Sn_{0.76}Ru$	Pb	1150	1		318	1005
YB_2	Y	2000	5		223	111
$YbRh_2Si_2$	Zn	1150	20		304	114
YVO_4	$LiVO_3$	1200	4	Part of a seeded study	134	1
YVO_4	V_2O_5	1450	10		148	193
$ZnCr_2O_4$	PbO–PbF_2–MoO_3	1220	1		54	607
$ZnGa_2O_4$	PbO–B_2O_3	1250	10	Other studies exist	171	131
ZnO	KOH–NaOH–LiOH	260	18		336	56
ZnO:*M*	KOH	600	10	M = Cr, Mn, Fe, Co	314	123
ZnS	$PbCl_2$	950	2	Temp. gradient; other studies	267	74
$ZnSiO_4$	$Pb_2ZnSi_2O_7$	1300		ZnF_2 also added	60	219
$ZnSiO_4$	Li_2MoO_4	1300	30		114	373
$ZrMo_2O_8$	Li_2MoO_4	750	3		404	100
$ZrMO_4$	Na_2O–MoO_3	1350	5	M = Si, Ge; also Zr=Hf	116	151
ZrO_2–R_2O_3	KF–$Na_2B_4O_7$	950	3	Also HfO_2–R_2O_3	94	287
ZrO_2–Y_2O_3	KF–$Na_2B_4O_7$	910	1	Evaporation	75	630
$ZrSiO_4$	PbO–PbF_2–MoO_3		5		43	336
$ZrSiO_4$	Li_2WO_4–WO_3	1300	2	Many dopants studied	125	431
ZrW_2O_8	WO_3	1300	10		212	167
ZrW_2O_8	WO_3	1280	4	Evaporation	343	115

[a]For alkali and alkali-earth oxides, corresponding carbonates are usually used as the starting material

Index

M. Tachibana, *Beginner's Guide to Flux Crystal Growth*, NIMS Monographs,
DOI 10.1007/978-4-431-56587-1

GPSR Compliance
The European Union's (EU) General Product Safety Regulation (GPSR) is a set of rules that requires consumer products to be safe and our obligations to ensure this.

If you have any concerns about our products, you can contact us on

ProductSafety@springernature.com

In case Publisher is established outside the EU, the EU authorized representative is:

Springer Nature Customer Service Center GmbH
Europaplatz 3
69115 Heidelberg, Germany

www.ingramcontent.com/pod-product-compliance
Ingram Content Group UK Ltd.
Pitfield, Milton Keynes, MK11 3LW, UK
UKHW021834270726
14058UKWH00001B/150

* 9 7 8 4 4 3 1 5 6 5 8 6 4 *